THE BATTLESHIPS OF THE IOWA CLASS

Missouri opens up with her main battery during exercise RimPac 90.
NARA

THE BATTLESHIPS OF THE *IOWA* CLASS

A DESIGN AND OPERATIONAL HISTORY

by Philippe Caresse

TRANSLATED BY BRUCE TAYLOR

NAVAL INSTITUTE PRESS
Annapolis, Maryland

Naval Institute Press
291 Wood Road
Annapolis, MD 21402

English edition © 2019 by the United States Naval Institute
Translation © 2019 by Bruce Taylor
French edition © 2015 by Lela Presse
All rights reserved.
The original edition was published under the title *Les Cuirassés de la classe « Iowa, »* Volumes 1 and 2.
The original French edition was published by Éditions Lela Presse, Le Vigen, France.

Library of Congress Cataloging-in-Publication Data
Names: Caresse, Philippe, date, author. | Taylor, Bruce, date, translator.
Title: The battleships of the Iowa class : a design and operational history / Philippe Caresse ; translated by Bruce Taylor.
Other titles: Cuirassés de la classe Iowa. English
Description: Annapolis, MD : Naval Institute Press, [2019] | Originally published: Les cuirassés de la classe Iowa. Le Vigen, France : Éditions Lela Presse. | Includes bibliographical references and index.
Identifiers: LCCN 2019007721 | ISBN 9781591145981 (hbk. : alk. paper)
Subjects: LCSH: Iowa Class (Battleships)—History. | Iowa Class (Battleships)—Design and construction. | World War, 1939–1945—Naval operations, American.
Classification: LCC V815.3 .C3713 2019 | DDC 359.8/3520973—dc23
LC record available at https://lccn.loc.gov/2019007721

♾Print editions meet the requirements of ANSI/NISO z39.48-1992 (Permanence of Paper).
Printed in the United States of America.

27 26 25 24 23 22 21 20 19   9 8 7 6 5 4 3 2 1
First printing

Maps by Chris Robinson.
All 3-D drawings by Stephan Draminski.

New Jersey off the coast of Provence, France, in 1984.
PRADIGNAC AND LEO

Contents

	Foreword	vii
	Preface	ix
1	The Genesis of the *Iowa*-Class Battleships	3
2	Technical Characteristics	18
3	Armament	46
4	Power and Propulsion	82
5	Shipboard Equipment	98
6	Paint and Camouflage	184
7	Commanding Officers	192
8	Battle Honors	198
9	Refitting and Conversion Projects	200
10	Previous Ships of the Name	202
11	The Sad End of the *Illinois* and *Kentucky*	209
12	Preservation and Legal Status	210
13	The Career of the Battleship USS *Iowa* (BB 61)	219
14	The Career of the Battleship USS *New Jersey* (BB 62)	304
15	The Career of the Battleship USS *Missouri* (BB 63)	382
16	The Career of the Battleship USS *Wisconsin* (BB 64)	457
	Sources	514
	Acknowledgments	515
	Index	516

PART I — Genesis and Technology

PART II — Service Careers

The engine power and fine hull lines of the *Iowa*s (in this case *Missouri*) gave the class a speed in excess of 33 knots, unmatched by any battleship in history.
USN

Foreword

The *Iowa*-class battleships were designed during the late 1930s, when there was great uncertainty as to what speed, firepower, and armor had been incorporated into potential enemies' new warship construction or in upgrades to their existing vessels. The United States had respected the various naval treaties limiting warship numbers, size, and firepower, but had the world's other navies done so as well? With limited intelligence as to the capabilities of the vessels of rival nations and concern for the unknown as war clouds gathered, the U.S. Navy set out to build a class of battleships that had commanding speed to hunt down and defeat any warship it encountered.

American battleships were traditionally heavily armored and gunned, but were slower than their contemporaries. The new *Iowa*-class design broke this mold and was designed with the greatest speed, armor, and firepower then available. With no expense spared as to their capabilities, the price tag swelled 60 percent beyond the preceding *North Carolina* and *South Dakota* classes. Congress was shocked at this price but funded their construction to support the needs of the two-ocean fleet required to fight the approaching world war. The capabilities of the *Iowa* class would ensure their longevity and relevance in several more American conflicts beyond World War II, including the Korean War, the Vietnam War, and the Cold War.

Six *Iowa*s were authorized and four were commissioned: *Iowa* (BB 61), *New Jersey* (BB 62), *Missouri* (BB 63), and *Wisconsin* (BB 64). The two uncommissioned units, *Illinois* (BB 65) and *Kentucky* (BB 66), were 22 percent and 73 percent complete before being scrapped. There were plans to complete *Kentucky* as a guided-missile battleship, fast oiler, and even a high-speed replenishment vessel, but none of these came to fruition. Her fate was decided with the removal of her bow to replace that of *Wisconsin* after the latter was in a collision with the destroyer *Eaton* in 1956. *Wisconsin* was thereafter jokingly known as the "Wistucky."

The century of the American battleship lasted from 1896 to 1995 and traces its origins to the ironclads of the American Civil War. The U.S. Navy ordered eighty-one capital ships, of which fifty-nine were completed: fifty-seven battleships and two battlecruisers (*Alaska* and *Guam*, albeit referred to as large cruisers by some). A total of twenty-two vessels (fourteen battleships and eight battlecruisers) were ordered but canceled before completion. Of those twenty-two vessels, the Washington Naval Treaty of 1922 canceled eleven (seven battleships and four battlecruisers). Over the entire ninety-nine years of the U.S. Navy battleship era, no battleships were sunk by enemy action at sea. *Utah*, *Oklahoma*, and *Arizona* were lost as a result of the attack on Pearl Harbor, while the armored cruiser *Maine* was destroyed by an internal explosion in Havana Harbor, Cuba, in 1898.

Of the fifty-seven battleships commissioned for the U.S. Navy, eight have survived as memorial museum warships on display in America. *Iowa* was the last battleship in the world to be saved and in September 2011 was awarded to the nonprofit Pacific Battleship Center to become the only vessel of her type on display on the West Coast of America. How very appropriate for *Iowa* to be permanently berthed in the Los Angeles/Long Beach harbor, which was once called "Battleship Country" and served as the homeport to several U.S. battleships, including *Iowa* during two of her three commissionings.

In October 2013, while working in the battleship *Iowa*'s archives, I was summoned topside to meet visiting author Philippe Caresse along with two gentlemen he was traveling with. I soon realized they were all fellow warship enthusiasts and took them on a special tour of *Iowa*'s many compartments. This meeting led Philippe and me to begin sharing and exchanging information on warships as well as many unique photographs from both our collections. After receiving copies of his many outstanding volumes on warships, I was thrilled to learn he had decided to write a book on the *Iowa* class, a topic that had never been fully covered for French readers and has now been translated into English.

Besides providing an excellent source of reference data and historical detail, this work is also wonderfully illustrated with highly accurate 3D drawings and an outstanding collection of photographs, many never previously published, making it a visual treat for model ship builders. It also documents, using profile and deck line drawings, the modifications made to the class during the ships' many years of service. The result is an engaging record of one of America's greatest warship classes, and I have no hesitation in recommending it to anyone with an interest in this truly imposing piece of our nation's naval heritage, now preserved in its entirety.

—David Way
Curator, Battleship *Iowa*
San Pedro, Port of Los Angeles
August 2018

New Jersey anchored off the French port of Villefranche-sur-Mer in April 1984.
PHILIPPE CARESSE

Preface

Lying at San Pedro, Philadelphia, Pearl Harbor, and Norfolk are the four greatest battleships still afloat, the result of the incredible feat by the United States of preserving the *Iowa*, *New Jersey*, *Missouri*, and *Wisconsin*, respectively, as museum ships. These four giants are in a remarkable state of preservation, making it possible for us to admire a super-dreadnought at close quarters, the ne plus ultra of battleship design.

The *Iowa*-class battleships had exceptionally long careers that have left their mark on naval history. Not only did they serve as Task Group flagships during the Pacific War, but they also saw action during the Korean War and one, *New Jersey*, was activated during the Vietnam War. During the 1980s *Iowa* and *New Jersey* were assigned to the Sixth Fleet in the Mediterranean to preserve U.S. interests during the war in Lebanon. *Iowa* appeared in the Persian Gulf in 1991 following the Iraqi invasion of Kuwait, and later that same year *Missouri* and *Wisconsin* shelled Iraqi positions with their 16-inch guns. These vessels were therefore intermittently operational from 1943 or 1944 until being finally withdrawn from service in the early 1990s, a record for larger vessels of the type.

With such longevity it is no surprise that these ships experienced a significant evolution in their armament and electronic equipment, so that during the war in Lebanon and the Gulf War the battleships were equipped with 20-mm Phalanx guns and Tomahawk and Harpoon missiles. The means of detection were also modernized with the installation of two dimensions of radar of different types.

The ships of the *Iowa* class were and remain major achievements of which the U.S. Navy can be proud. They were capable of meeting any battleship in single combat. Neither the *Bismarck*, *Littorio*, nor *Nagato* class could rival them either in protection, firepower, or speed, with only the *Yamato* class standing comparison with these vessels. It is, however, meaningless to speculate on the outcome of any engagement. Too many factors are in play, notably the human element, since these steel monsters could not operate at full efficiency without the skill and dedication of trained crews, a point frequently overlooked.

The *Iowa* class also suffered the hard knocks of war. In 1944 *Iowa* was struck by a shell off Mili Atoll and *Missouri* by kamikaze in April 1945. The latter entered history by receiving the delegation that signed the Japanese Instrument of Surrender on her decks in September of that year. *New Jersey* and *Wisconsin* were damaged during the Korean War, but the most serious event occurred on board *Iowa* in April 1989 with the explosion in Turret 2, a tragedy that claimed forty-seven lives.

From a purely personal standpoint, the author's particular attachment is to *New Jersey*, simply on the strength of having witnessed her at sea while serving in Olifant 19, the French military operations conducted during the Lebanon War in 1984. Embarked as a crewman in the French destroyer *d'Estrées*, I would sometimes hear announcements on the loudspeaker that she was in sight. The firing of the 16-inch guns was extremely impressive and at that time I never imagined the day might come when I could visit these mastodons in repose.

When that day finally came nearly thirty years later, the visit in question was not to *New Jersey* but to *Iowa*, whose crew, volunteers, and curator not only gave me the warmest of welcomes but let me tour the battleship from top to bottom, and over the course of five unforgettable hours I had a chance to view the ship in the smallest detail.

In the pages that follow I have attempted to provide a technical and historical account of this great class of battleships, and if you have not done so already I would urge you to lose no time in visiting these marvels of naval technology from another age.

—Philippe Caresse
Théoule-sur-Mer, Alpes-Maritimes, France
August 2018

New Jersey seen lying at Camden beyond the bows of the protected cruiser *Olympia*, herself berthed at Philadelphia on the opposite bank of the Delaware River.
PHILIPPE CARESSE

PART I
Genesis and Technology

ONE

The Genesis of the *Iowa*-Class Battleships

With the expansion of U.S. territory following the Spanish-American War of 1898, it became clear that the United States of America would have to equip itself with a large navy capable of operating over much of the Pacific. During the first decades of the twentieth century it was accepted that the most important and the most powerful vessel in any fleet was the battleship. The U.S. Navy therefore equipped itself with a number of such vessels, the first units of which had questionable military and seagoing capabilities.

After the completion of the revolutionary British battleship *Dreadnought* in 1906, the U.S. Navy commissioned the all-big-gun battleships of the *South Carolina* class, albeit designed before the construction of HMS *Dreadnought*. The caliber of the main armament was 12 inches mounted in twin turrets, with propulsion provided by boilers and triple-expansion engines. The silhouette of these vessels would also be defined for many years by cage masts, two stacks, and a secondary armament disposed in casemates. The same 12-inch caliber was retained in the *Delaware*, *Florida*, and *Wyoming* classes, but the engine spaces now included turbines, permitting a more flexible and efficient use of steam power. Guns of 14-inch caliber were adopted for the *New York*, *Nevada*, *Pennsylvania*, and *New Mexico* classes, but speed remained at a relatively modest maximum of 21 knots. With these sixteen dreadnoughts, however, the United States emerged from World War I as one of the great naval powers, second only to the Royal Navy.

By the end of World War I the Japanese had commissioned the four battlecruisers of the *Kongō* class, the first unit of which was both designed and built in Britain. Armed with eight 14-inch guns, they were capable of 27 knots and had a range of eight thousand miles. Reclassified as battleships between the wars, together with the British battlecruisers these were the only capital ships capable of escorting that newcomer to the world's fleets, the aircraft carrier. Indeed, Japan was among the pioneers in the development of these new vessels, commissioning the *Hosho* in 1922, which was capable of 25 knots. She was followed in 1927 by the *Akagi* class, capable of 31 knots.

Returning to battleships, by 1918 the Imperial Japanese Navy had also commissioned the powerfully armed *Fusō* and *Ise* classes, followed in 1920 by *Nagato* and *Mutsu*, mounting eight 16-inch guns and capable of 27 knots on 34,000 tons. Many warships of all types were either under construction or planned, and a new naval arms race seemed to be in full swing.

In an effort to avert this situation, the United States decided to call a conference of the five great naval powers. The Washington Naval Conference began in November 1921 and concluded with a treaty in February 1922 to which the United States, Great Britain, Japan, France, and Italy were signatories. Limitations were placed on the number of vessels, their displacement and armament, amounting to a ratio of 5:5:3:1.75:1.75, respectively, with respect to capital ship tonnage. Moreover, with a few exceptions, a moratorium of ten years was imposed on the construction of battleships and battlecruisers with immediate effect. In addition, no vessel could be replaced before it had been in service for twenty years.

The global tonnage was established as follows:

	BATTLESHIPS AND BATTLECRUISERS	AIRCRAFT CARRIERS
Great Britain	525,000 tons	135,000 tons
United States	525,000 tons	135,000 tons
Japan	345,000 tons	81,000 tons
France	175,000 tons	60,000 tons
Italy	175,000 tons	60,000 tons

The main features of these vessels were to be as follows:

	MAXIMUM DISPLACEMENT	MAXIMUM CALIBER
Battleships and Battlecruisers	35,000 tons	16-inch
Aircraft Carriers	27,000 tons	8-inch
Cruisers	10,000 tons	8-inch

With *Nagato* and *Mutsu* having recently joined the fleet, Japan initially seemed not to oppose the 5:5:3:1.75:1.75 ratio, and the consequences of this treaty for the United States was a halt in the construction of battleships that lasted nearly twenty years. Completion of the six *South Dakota*–class battleships was canceled in 1922 together with the six battlecruisers of the *Lexington* class, but *Lexington* and *Saratoga* were converted to aircraft carriers and entered service in 1927. Although the majority of its dreadnoughts were rebuilt between the wars, the U.S. Navy would inevitably be short of fast modern battleships.

The battlecruiser *Kongō*, built in Britain for the Imperial Japanese Navy. She was armed with eight 14-inch guns and was capable of 27 knots.
USN

The formidable Japanese battleship *Nagato*, mounting eight 16-inch guns and capable of 27 knots.
USN

A 16-inch barrel intended for mounting in the battleships of the *South Dakota* class, the construction of which was suspended in 1922. This example is preserved at the Washington Navy Yard.
PHILIPPE CARESSE

The *Delaware*-class battleship *North Dakota* (BB 29), seen here in dry dock c. 1917–19.
USN

The 12-inch battery of the USS *Florida* (BB 30).
USN

The USS *Wyoming* (BB 32) during her speed trials in 1912.
USN

The dreadnought USS *New York* (BB 34) nearing completion at the New York Naval Shipyard.
USN

The *Nevada*-class battleship *Oklahoma* (BB 37) during a gunnery exercise with her 14-inch armament.
USN

The ill-fated *Pennsylvania*-class battleship USS *Arizona* (BB 39) seen in 1919.
USN

The USS *New Mexico* (BB 40) at full speed.
USN

South Dakota (BB 49) seen at the New York Naval Shipyard on March 10, 1922. A victim of the Washington Treaty, she was eventually scrapped incomplete.
NARA

The carriers *Lexington* (CV 2) and *Saratoga* (CV 3) seen in February 1933. They were originally designed as battlecruisers.
USN

The battleship USS *North Carolina* (BB 55) in December 1944.
USN

The Washington Treaty was followed in 1927 by the Geneva Naval Conference, which failed in its attempt to limit the number and features of cruisers, destroyers, and submarines. Agreement was reached on this at the London Naval Conference on April 22, 1930, but the more drastic restrictions proposed at the Second London Naval Conference in 1936 were regarded by the Japanese delegation as a threat to the empire, and it withdrew over a provision limiting vessels to guns of 14-inch caliber. There was, however, an escalator clause imposed by the U.S. negotiators should one of the signatories to the Washington Treaty of 1922 refuse to abide by that limit. The world's naval powers were on the road to war.

In response to the pressing need for a modern and homogeneous fleet, between May and July 1935 the General Board of the United States Navy reviewed designs for a new class of battleships (eventually the *North Carolina* class). The choice was restricted either to a well-armed and well-protected vessel at the expense of speed, or one capable of escorting aircraft carriers and matching the speed of the Japanese battleships, albeit with reduced armament and armor. Having studied thirty-five designs labeled A to M, the General Board settled in November 1935 on a vessel capable of 30 knots and mounting nine 14-inch guns on 35,000 tons. Secretary of the Navy Claude A. Swanson subsequently ordered a modified version with four 14-inch turrets capable of 27 knots, although solutions were limited by a displacement of 35,000 tons. By August 1936, design XVI-C, consisting of eleven 14-inch guns and a speed of 30 knots had received the approval of the committee, but ran into opposition from Adm. Joseph Reeves, who not only disapproved of the design on grounds of cost but regarded its speed as insufficient to escort the U.S. Navy's new 33-knot aircraft carriers.

Design XVI-C was therefore modified with increased protection and the possibility of exchanging the 14-inch turret with a 16-inch version in the event of any breach of the Second London Naval Treaty. On March 27, 1937, Japan refused to restrict itself to the imposed limit, resulting in the announcement of the *North Carolina* class in what turned out to be almost its final form. President Roosevelt came under great political pressure, however, stating that "I am not willing that the United States be the first naval power to adopt the 16-inch gun. [. . .] Because of the international importance of the United States not being the first to change the principles laid down in the Washington and London Treaties, it seems to me that the plans for the two new battleships should contemplate the [. . .] 14-inch gun." However, Admiral Reeves spared no effort to secure the higher caliber, and on June 24 the construction of *North Carolina* (BB 55) and *Washington* (BB 56) was voted into the 1937 budget. On July 10, Roosevelt duly authorized mounting of a 16-inch/45-caliber armament, but it was recognized that these two battleships, displacing 45,000 tons at full load, would have considerable difficulty reaching 27 knots.

The battleship USS *Washington* (BB 56) off Port Angeles, Washington, on April 29, 1944.
USN

Naval architects at work on new battleship designs.
USN

The battleship USS *South Dakota* (BB 57), nameship of her class. She displaced 45,200 tons, was capable of 27 knots, and mounted nine 16-inch guns.
USN

The budget for fiscal year 1938 therefore saw the appearance of two battleships supplementary to the *North Carolina* class. Work on the design began in March 1937 and construction was approved by Secretary of the Navy Swanson on January 4, 1938. But the Chief of Naval Operations, Adm. William Standley, had already ordered a new study incorporating 16-inch guns from the outset together with increased protection. The internal arrangements were redesigned, particularly with respect to the engine plant, which permitted reduction to a single stack. Meanwhile, the deterioration in diplomatic relations in Europe as well as in Asia resulted in authorization for the construction of two additional vessels in this class on June 25 that year. With a full-load displacement of 45,200 tons, a speed of 27 knots, and an armament of nine 16-inch/45-caliber guns, the *South Dakota*s are widely

The battleships USS *Indiana* (BB 58) and *Massachusetts* (BB 59) and the heavy cruisers USS *Chicago* (CA 136) and *Quincy* (CA 71) steaming in line ahead.
NARA

regarded as the finest battleship design produced in the context of the Washington Treaty, albeit with significantly increased tonnage. The class consisted of *South Dakota* (BB 57), *Indiana* (BB 58), *Massachusetts* (BB 59), and *Alabama* (BB 60).

At this time the U.S. Navy had a number of aircraft carriers in service or under construction capable of speeds approaching or in excess of 30 knots, including *Ranger* and *Wasp*, which were the slowest at 29.5 knots, while *Lexington*, *Saratoga*, *Yorktown*, and *Enterprise* were all capable of 32 to 33 knots. It therefore seemed imperative to equip the fleet with capital ships capable of maintaining that speed, quite aside from the reports that appeared in the Italian press in November 1937 regarding Japanese plans to lay down a class of 46,000-ton vessels armed with 16-inch guns. Although these claims were as yet unconfirmed, the U.S. Navy decided to adopt the new 16-inch/50-caliber gun in future construction. Meanwhile, the Japanese invasion of China in July 1937 and the German annexation of Austria in March 1938 brought the prospect of war that much closer, and on May 17 of that year the United States promulgated the Navy Construction Act (known as the Second Vinson Act) providing for a 20 percent increase in the size of the fleet.

The first work on this new generation of battleships dates back to early 1938, when the chairman of the General Board, Adm. Thomas C. Hart, took over direction of the design studies for future vessels. In the meantime, the Bureau of Construction and Repair had on January 17, 1938, begun working on a cruiser killer concept under Capt. Allan J. Chantry armed with twelve 16-inch guns and twenty 6-inch guns and capable of making 35 knots on 51,760 tons. The scheme of protection, however, could only withstand 8-inch shells. By the end of January three designs were available for review: Design A, of 60,000 tons with twelve 16-inch guns in triple turrets generating 277,000 shp for 32.5 knots; Design B, of 53,500 tons mounting nine 16-inch guns and capable of the same speed; and Design C, which shared the same specifications as Design B although with 300,000 shp yielding a best speed of 35 knots.

Nonetheless, in March 1938 the General Board followed the recommendations of the Battleship Design Advisory Board, composed of the naval architects William Hovgaard, Joseph W. Powell, John Metten, and the head of the Bureau of Ordnance, Adm. Joseph Strauss, which advocated the adoption of a wholly new design. By this time U.S. intelligence services had learned of the approximate specifications of what was to become the *Yamato* class, consisting of a standard displacement of 65,000 tons, an armament of 18-inch guns and a speed of 26 knots.

With the escalator clause more pertinent than ever, the displacement of the latest vessels was set at 45,000 tons. It was by now out of the question to equip the new vessels with 18-inch/47-caliber guns, but a new design of 16-inch shell was reckoned to make good the difference in size. The Mk-7 50-caliber 16-inch gun could fire a 2,700-pound AP shell with a penetration equal to the Japanese 18-inch model. On April 21 the design for the new 16-inch guns and their turrets was adopted.

On May 17, 1938, Congress received and approved the construction program for the first two units of the *Iowa* class, BB 61 and BB 62, with the final plans signed off by the New York–based design bureau of Gibbs & Cox. Two weeks later, on June 2, the Bureau of Construction and Repair submitted what would become the designs for the *Iowa* class to the General Board, and on June 9 the Secretary of the Navy received the dossier on his desk. The design was based on the hull of the *South Dakota* class but lengthened by no less than 164 feet to attain a speed in excess of 30 knots. Beam remained essentially the same in order to permit passage through the Panama Canal, and the classic boiler/turbine arrangement was retained. The particulars were as follows:

Standard displacement	44,560 tons
Full load displacement	55,710 tons
Length	860 feet at the waterline
Beam	108 feet 3 inches
Armament	9 × 16-inch guns
	20 × 5-inch guns
	12 × 1.1-inch guns in quadruple mounts
Protection	Resistant to 16-inch/50 shells at ranges from 21,750–32,000 yards
Installed power	200,000 shp
Speed	33 knots
Range	15,000 nm at 15 knots

The Panamax specifications permitted no greater beam since the class had clearance of only a foot on either side when traversing the canal.

The battleship *Yamato* fitting out at Kure on September 20, 1941.
USN

The Genesis of the *Iowa*-Class Battleships   15

USS *Missouri* (BB 63) and the *Essex*-class carrier USS *Bennington* (CV 20) fitting out at the New York Naval Shipyard.
USN

This ambitious program was intended to put six battleships into service, including the USS *Iowa* (BB 61), *New Jersey* (BB 62), *Missouri* (BB 63), *Wisconsin* (BB 64), *Illinois* (BB 65), and *Kentucky* (BB 66). *Iowa* and *Missouri* were built at the New York Naval Shipyard, which was also responsible for finalizing and sharing the plans, of which there were 6,000 for the hull, 700 for the armament, 2,400 for the electrical arrangements, and 2,700 for the engine plant and related systems. Construction required a total of more than 20,000 plans for all sections. *New Jersey* and *Wisconsin* were built at the Philadelphia Naval Shipyard.

The contract for the construction of *Iowa* herself was signed on July 1, 1939, and that for *New Jersey* ratified on the 4th. The construction of *Missouri* and *Wisconsin* was approved by Congress on July 6, 1939, and the yards officially received their orders on June 12, 1940. The average construction cost of a unit of the class was $125 million.

As discussed below, the *Illinois* and *Kentucky* shared very different fates from their sisters. In the context of the war unfolding in Europe, Secretary of the Navy Frank Knox authorized construction on September 9, 1940, but the two vessels had not been completed by the end of the conflict and the hulls were retained until being scrapped in 1958.

The Builders

The New York Naval Shipyard was founded by the Jackson brothers in Brooklyn on the south bank of the East River facing the eastern shore of the island of Manhattan in 1781. It was recognized as such by Congress in 1800 and began building vessels for the U.S. Navy in 1817. In 1862 it completed the revolutionary *Monitor*, followed in due course by a series of capital ships including the armored cruiser (sometimes called a battleship) *Maine*, the predreadnoughts *Connecticut* (BB 18) and *New Hampshire* (BB 25), and eventually by the dreadnoughts *Arizona* (BB 39), *New Mexico* (BB 40), *Idaho* (BB 42), and *Tennessee* (BB 43). To these were added numerous aircraft carriers (including of the *Essex* class), heavy and light cruisers that added luster to the reputation of this famous yard, which closed in 1966.

The Philadelphia Naval Shipyard.
USN

The Philadelphia Naval Shipyard was established on the banks of the Delaware River in 1776 and became an official U.S. Navy yard in 1801. Having built vessels of all types between 1850 and 1870, the yard was used as a major maintenance and repair facility during World War I, employing 12,000 workers. During the 1930s it was responsible for the construction of the heavy cruisers *Minneapolis* (CA 36) and *Wichita* (CA 45), the light cruiser *Philadelphia* (CL 41), and numerous destroyers. These were followed by the battleship *Washington* (BB 56) together with units of the *Essex* class and a number of cruisers completed at League Island. By the end of World War II the yard employed some 47,000 workers but ceased naval construction in 1995 and has now been largely redeveloped.

The Japanese carrier *Akagi*, capable of steaming in excess of 31 knots.
USN

Iowa, the first unit in a class of battleships destined to make naval history.
PRADIGNAC AND LEO

The USS *Alabama* (BB 60), now preserved at Battleship Park, Mobile, Alabama.
USN

TWO
Technical Characteristics

The hull of the *Iowa* class was divided into 216 frames and incorporated 1,135,000 rivets together with 800 miles in lines of welding. The freeboard was 36 feet forward and 22 feet aft. For the first time in U.S. battleship construction a bulbous bow was fitted to improve seakeeping qualities. A bulkhead was fitted on each side of the hull under deck 3 that extended to the triple bottom. The loading of 61.55 tons would cause the ship to settle 0.4 inches in the water. The hull surface was 113,592 square feet.

Iowa 1944

DIMENSIONS, DISPLACEMENT, AND PROTECTION

DIMENSIONS	IOWA	NEW JERSEY	MISSOURI	WISCONSIN
Overall length	887 ft 3 in	887 ft 7 in	887 ft 3 in	887 ft 3 in
Waterline length	859 ft 6 in	859 ft 10 in	860 ft	860 ft
Beam	108 ft 2 in	108 ft 1 in	108 ft 2 in	108 ft 2 in
Draft	28 ft 10 in at 43,875 t	38 ft at 57,216 t	37 ft 9 in at 57,540 t	38 ft at 57,216 t
Draft	37 ft 2 in at 59,331 t	37 ft 8.75 in at 57,500 t	37 ft 8.75 in at 57,500 t	37 ft 8.75 in at 57,500 t
DISPLACEMENT				
Displacement (design)	54,889 t	54,889 t	54,889 t	54,889 t
Normal displacement (1943)	43,875 t	43,944 t		
Standard displacement (1943)	49,202 t	45,649 t		
Full load displacement (1943)	57,540 t	57,216 t		
Full load displacement (1945)			57,540 t	57,216 t
Full load displacement (1983)		57,353 t		
PROTECTION				
Armor belt—center	12.2 in			
upper	7.2 in			
lower	4 in			
Bulkheads	Iowa & New Jersey 11.3 in; Missouri & Wisconsin 14.5 in			
Upper deck (bomb deck)	1.5 in			
Armor deck	4.75 in + 1.25 in			
Splinter deck	0.62 in			
Deck 3	0.51–0.62 in			

Source: Sumrall, Iowa Class Battleships, 156.

METACENTRIC HEIGHT

IOWA	NEW JERSEY
43,875 t = 4.88 ft	43,944 t = 5.15 ft
55,424 t = 7.68 ft	55,649 t = 7.95 ft
57,540 t = 8.40 ft	57,216 t = 8.57 ft

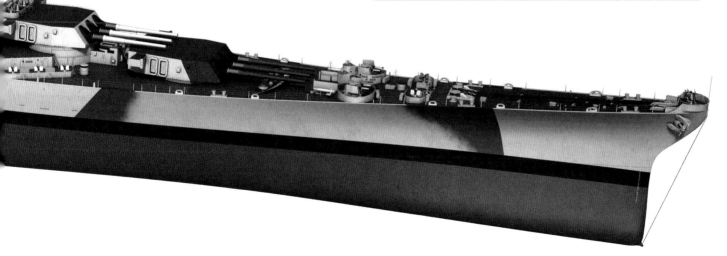

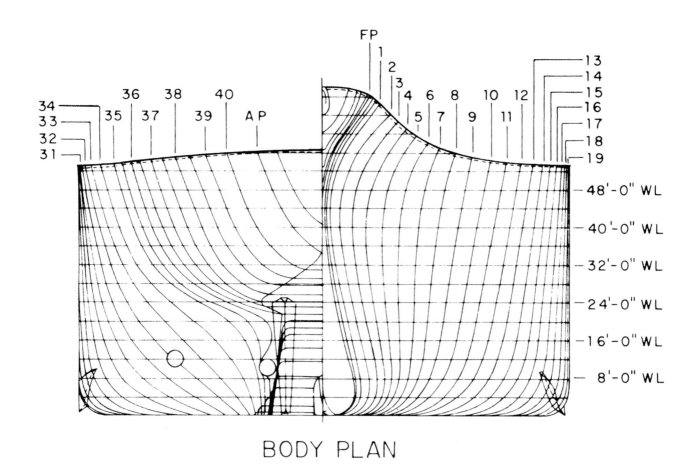

BODY PLAN

Rear Adm. Clark H. Woodward laying *Iowa*'s keel on June 27, 1940.
USN

The keel plate of *New Jersey* laid on September 16, 1940.
USN

Installation of *Wisconsin*'s watertight bulkheads in 1943.
USN

Missouri under construction in 1942; the boilers are already in place.
USN

Wisconsin under construction at the Philadelphia Naval Shipyard on January 12, 1943. Note the barbette of turret 3 in the foreground.
USN

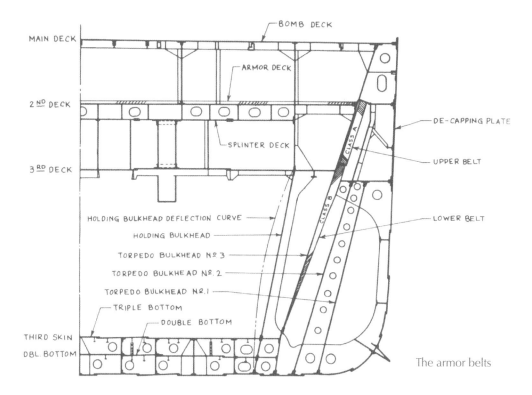

The armor belts

Armor

- The armor belt began at turret 1 and extended to turret 3. It had a length of 462 feet 7 inches and was inclined at 19 degrees. The upper part of this belt consisted of 1,085.79 tons of Class A (face hardened) steel, with the lower part consisting of 1,556.10 tons of Class B (homogenous) steel.
- There were three continuous decks. The main deck (bomb deck) was intended to detonate the shells and thereby facilitate the resistance of the main horizontal armor located below on the second deck, which together with the splinter deck extending from turret 2 to turret 3 protected the engine room and fireroom spaces, magazines, battery plot compartments, and the transmitter compartment. Within the armored citadel, deck 3 consisted of a tunnel running above the engine spaces and accessible from either the port or starboard 5-inch magazines and various other compartments. Abaft the citadel deck 3 provided protection for the steering compartments.
- Decks 2 and 3 were protected by 4,254.98 tons of Class B armor.
- Lying under the main deck were three platform decks from the bow to the forward armored bulkhead, and two platform decks from the stern to the after armored bulkhead.
- The upper decks, originally decked in teak, had an area of 53,820 square feet.
- Underwater protection consisted of a triple bottom with a design resistance to 700 pounds of explosive. Some 60 percent of the submerged area was protected.
- The forward and after armored bulkheads of *Iowa* were 11.3 inches in the upper section and tapering to 8.5 inches in the lower, being increased to 14.5 inches in tapering to 11.7 inches in the other units in consequence of the lifting of the restrictions of the Washington Naval Treaty.
- The steering gear at frames 189 and 203 was protected by armor varying from 5 or 6 inches to 13.5 inches.
- The displacement of the hull before fitting of the armor was 13,500 tons—28.9 percent.
- Total armor amounted to 18,700 tons—41.6 percent—supplied by Bethlehem Steel of Bethlehem, Pennsylvania, and Luken Steel of Coatesville, Pennsylvania.

Iowa's longitudinal internal passageway known as "Broadway."
NARA

The teak on *New Jersey*'s forecastle showing the wear of seventy-two years of service.
PHILIPPE CARESSE

Missouri's hull presents a wall of steel as she lies in dry dock on July 23, 1944.
USN

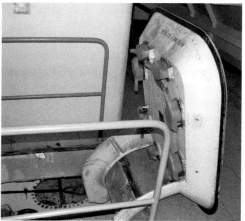

A hatch giving access below the armor deck protecting the matériel and equipment vital to the survival of the ship.
PHILIPPE CARESSE

Missouri in 1991.
J. PRADIGNAC

Wisconsin at sea in 1998.
J. PRADIGNAC

USS *Iowa*

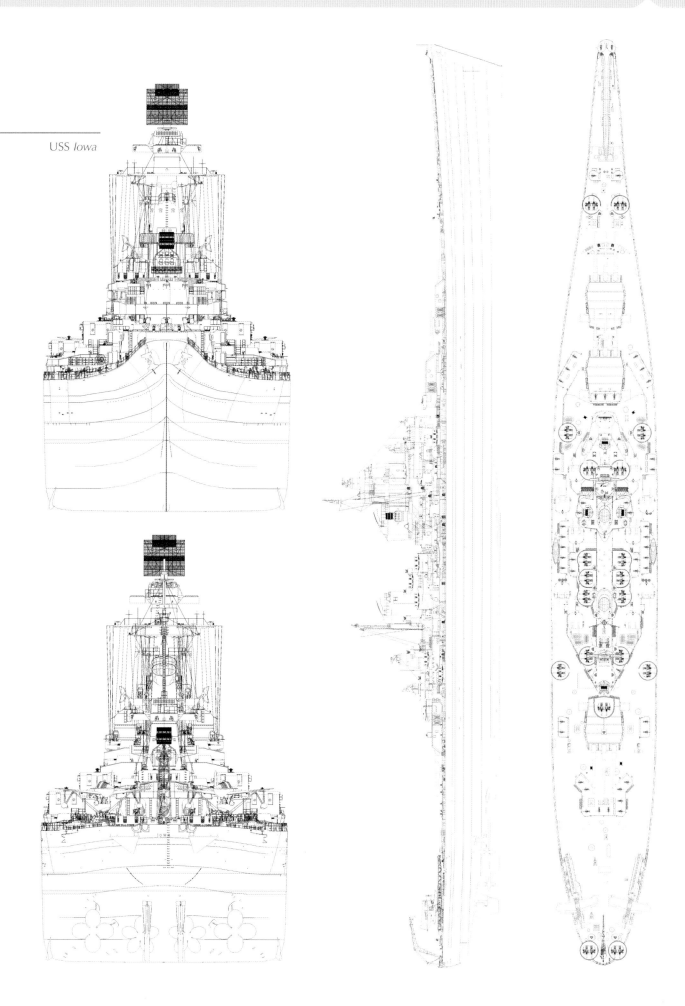

USS *Iowa* in 1944

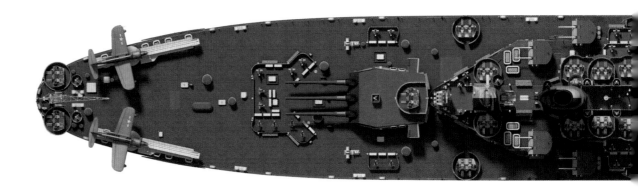

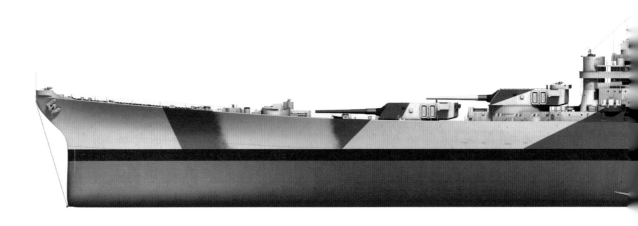

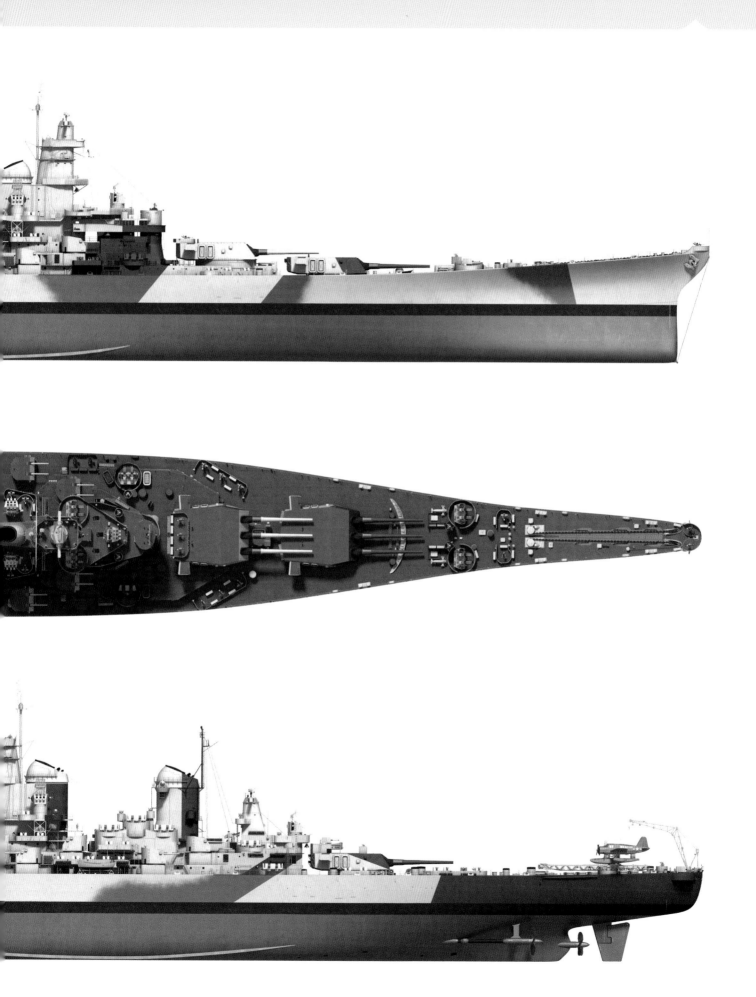

USS *Iowa*

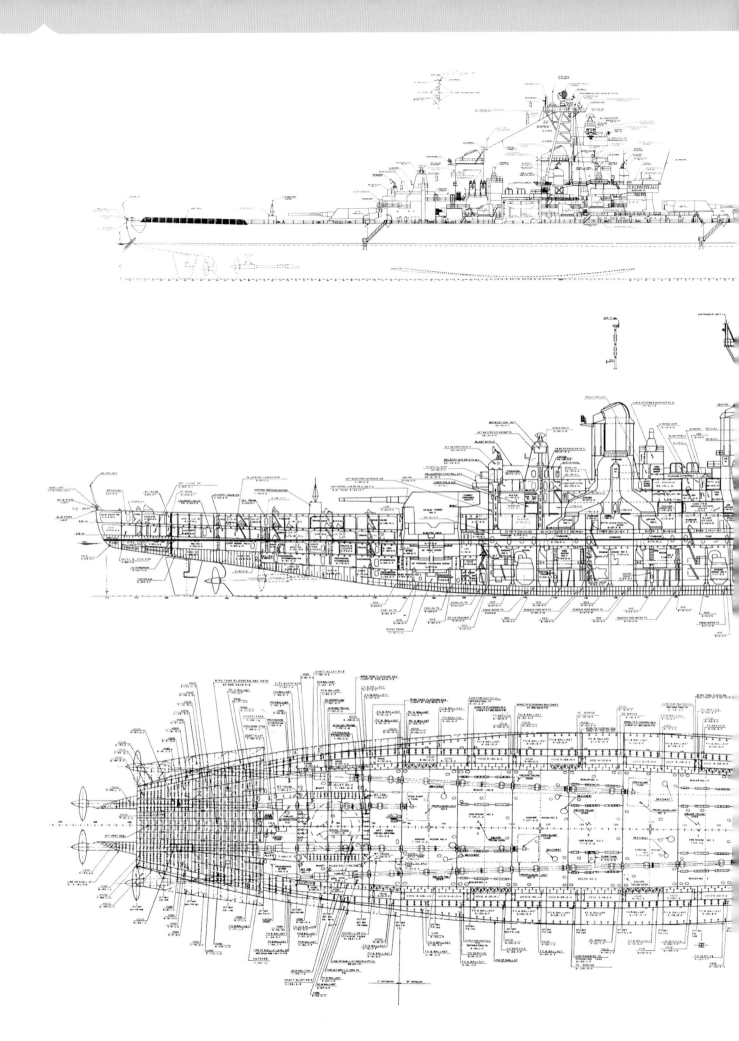

USS *Missouri*

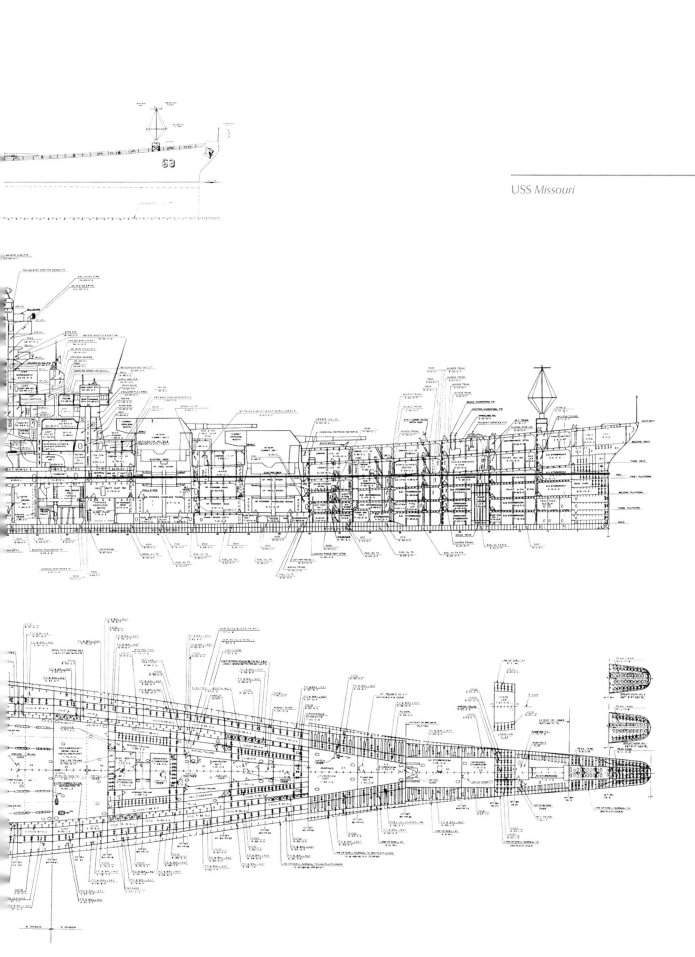

USS *Missouri* in 1945

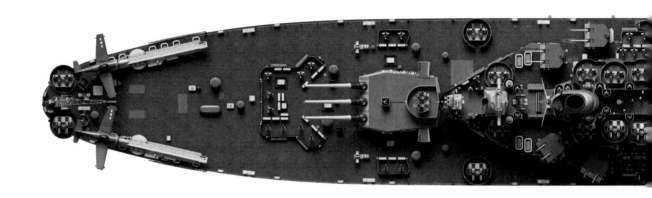
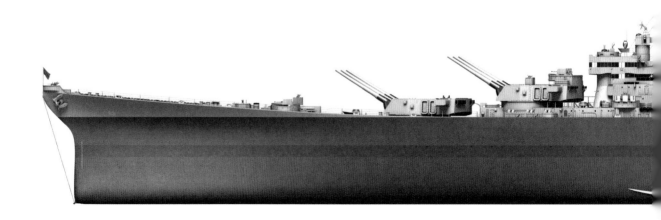

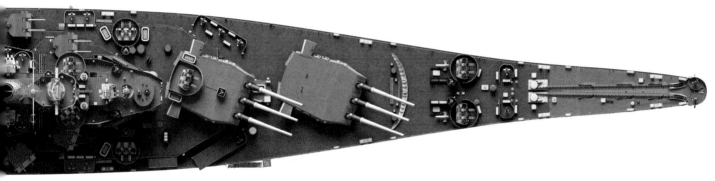

USS *Missouri*

USS *New Jersey* in 1983

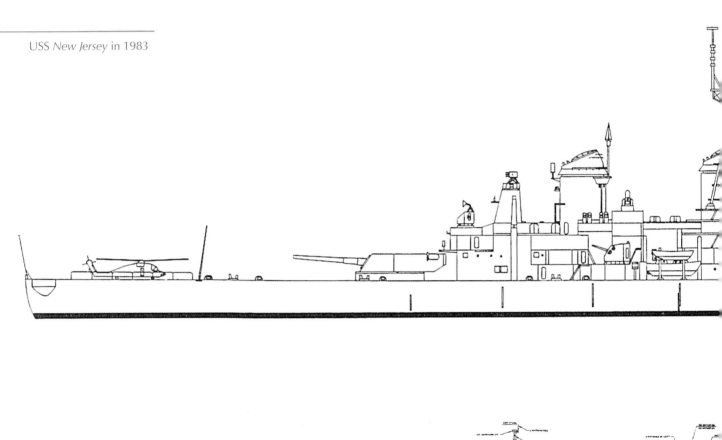
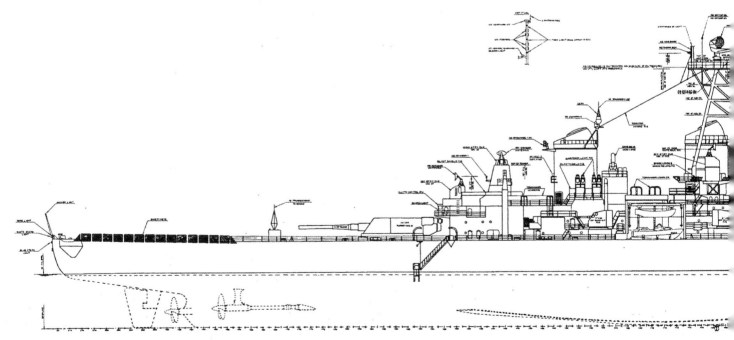

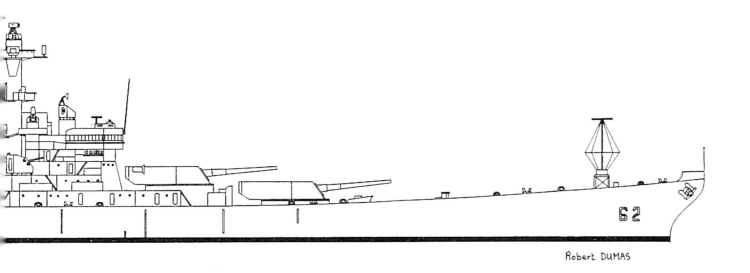

Robert DUMAS

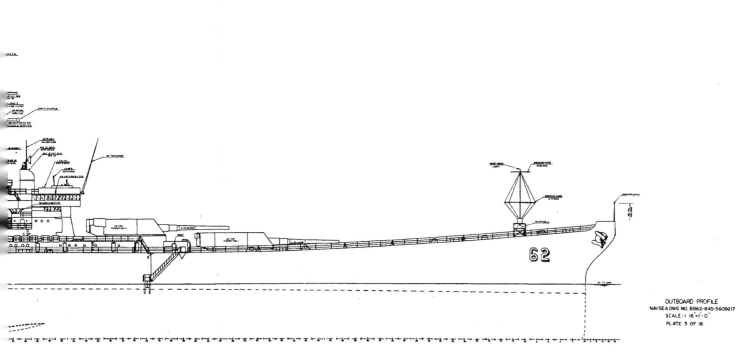

OUTBOARD PROFILE
NAVSEA DWG NO. BB62-845-5609217
SCALE: 1 16"=1'-0"
PLATE 3 OF 18

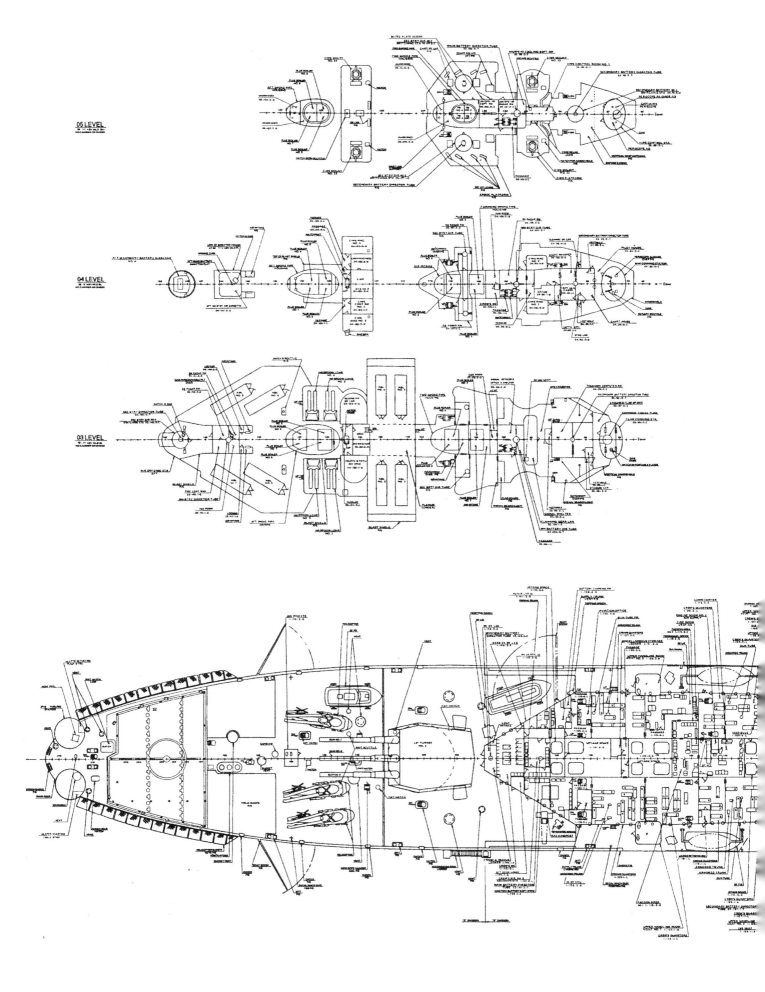

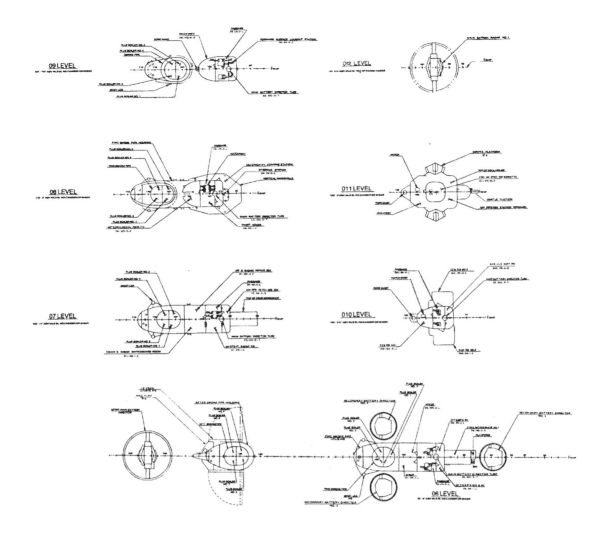
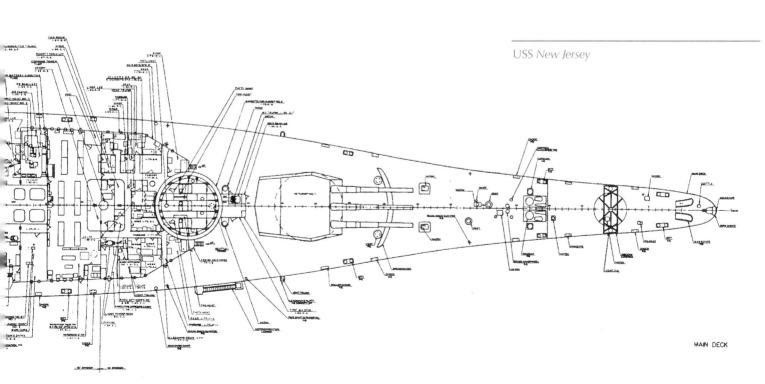

USS New Jersey

MAIN DECK

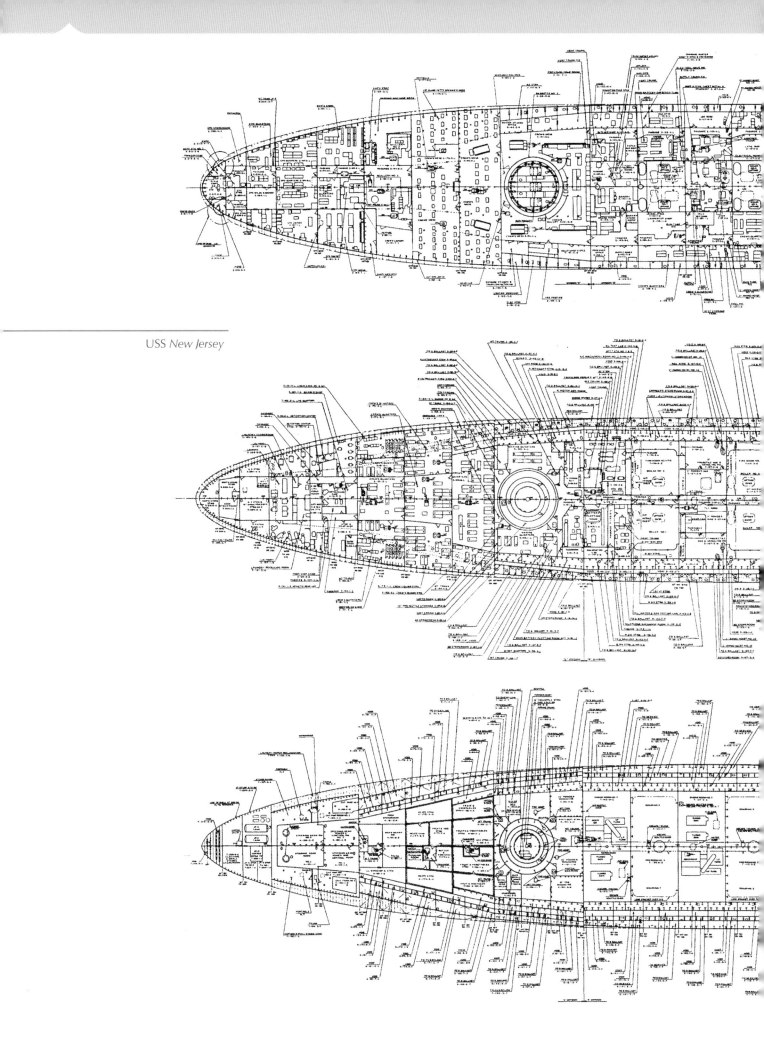

USS *New Jersey*

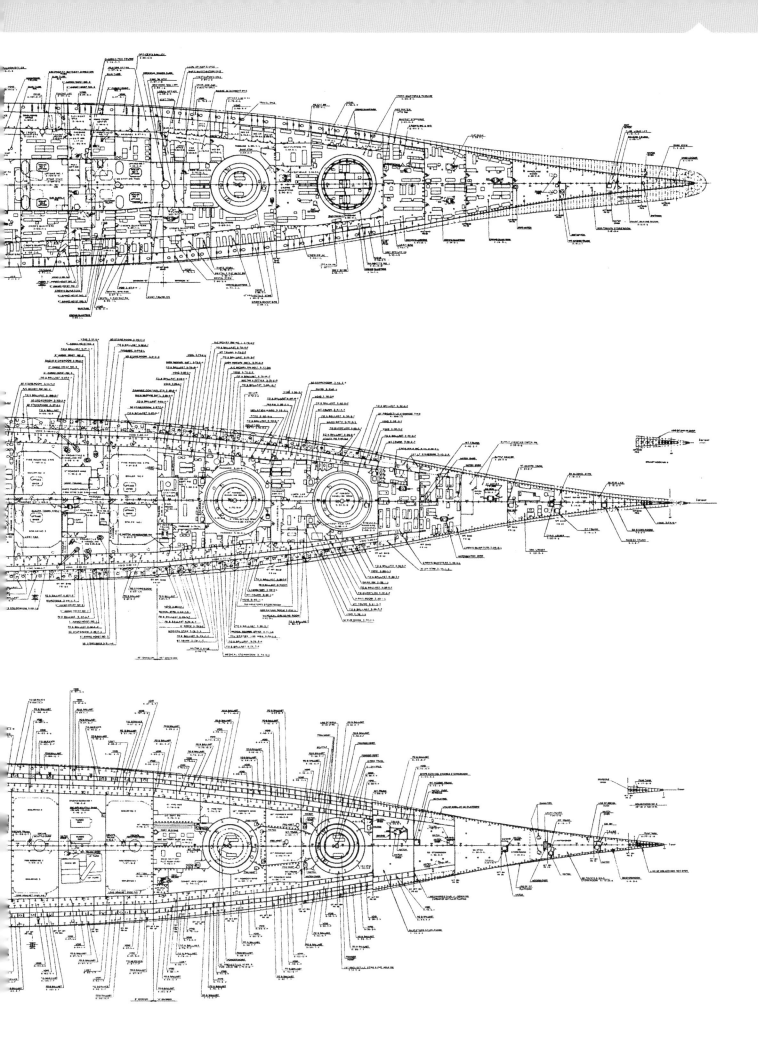

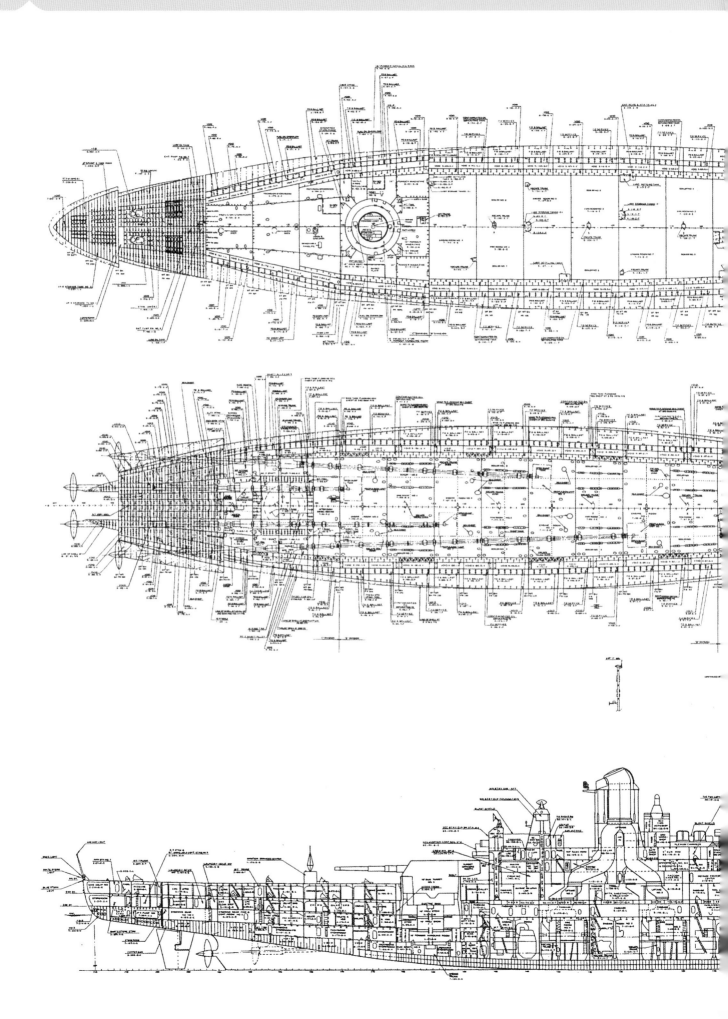

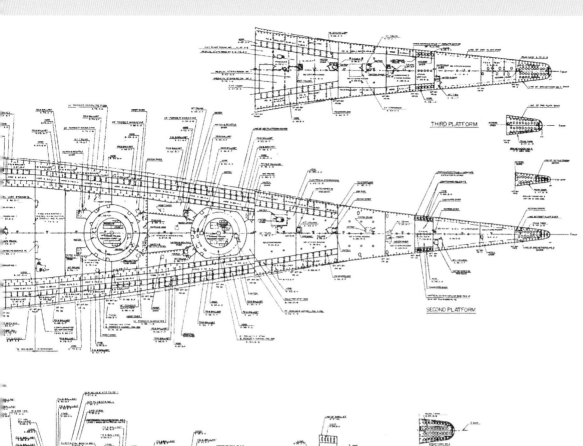
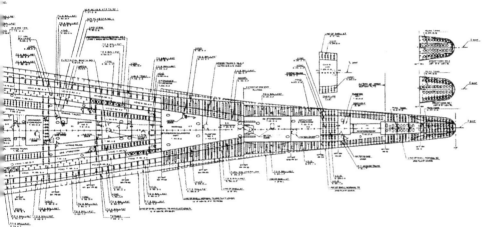

USS *New Jersey*

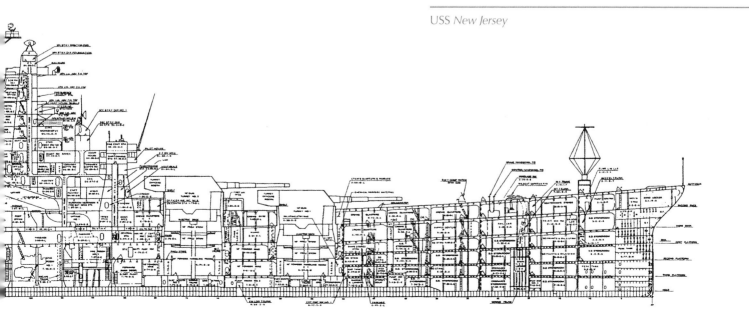

THREE

Armament

The armament of the class amounted to 10,800 tons (24 percent), and some 2,592 tons of ammunition was embarked in 1943. The weapon fit was subject to repeated modification, as follows:

ARMAMENT

1945 9 × 16-inch in 3 three-gun Mk-7 turrets
20 × 5-inch guns in 10 twin Mk-28 Mod 2 mounts
80 (*Iowa*: 76) × 40-mm in 20 quadruple Mk-2 mounts
49 (*Iowa*: 52) × 20-mm in single Oerlikon mounts

1951 9 × 16-inch in 3 three-gun Mk-7 turrets
20 × 5-inch guns in 10 twin Mk-28 Mod 2 mounts
76 (*New Jersey*: 72; *Wisconsin*: 66) × 40-mm in 20 quadruple Mk-2 mounts

1955 9 × 16-inch in 3 three-gun Mk-7 turrets
20 × 5-inch guns in 10 twin Mk-28 Mod 2 mounts
80 × 40-mm in 20 quadruple Mk-2 mounts

1968 (New Jersey)
9 × 16-inch in 3 three-gun Mk-7 turrets
20 × 5-inch guns in 10 twin Mk-28 Mod 2 mounts
4 × 5-inch Zuni Folding-Fin Aircraft Rockets (FFAR) (for launching chaff)

1982–88 9 × 16-inch in 3 three-gun Mk-7 turrets
12 × 5-inch guns in 6 twin Mk-28 Mod 2 mounts
4 × 20-mm Phalanx Mk-15 Block 0 CIWS
32 × BGM-109 Tomahawk cruise missiles
16 × RGM-84 Harpoon missiles
8 × 6 Mk-36 SRBOC launchers

Source: Sumrall, Iowa *Class Battleships*, 154, and David Way.

The 16-Inch Main Armament

The *Iowa*-class battleships were each equipped with three 16-inch Mk-7 turrets, two forward and one aft. The turrets were referred to as three-gun turrets instead of triple turrets, each gun having its own separate firing chamber. Each turret weighed 1,850 tons and each gun 167 tons. The cost per turret, less the guns, was $1,400,000.

Turret 1 consisted of seven decks (416 tons), turret 2 of twelve decks (770 tons), and turret 3 of seven decks (543 tons). The turret plating was supplied by the Charleston Naval Ordnance Plant of South Charleston, West Virginia. It was supplied with 390 shells, turret 2 with 460 shells, and turret 3 with 370 shells. Two types of shell were initially embarked, the AP Mk-8 (armor-piercing) of 2,700 pounds and the HC Mk-13 (high-capacity) of 1,900 pounds (costing $1,352 each in 1952). A bombardment of the Japanese coast in 1945 was reckoned to cost $74,255 per minute.

In 1956, *Iowa*, *New Jersey*, and *Wisconsin* were modified to receive nine Mk-23 Katie shells with 15- to 20-kiloton nuclear charges weighing 1,900 pounds each in the magazine of turret 2.

- The rate of fire of the guns was advertised as a shell every thirty seconds, but in practice it was approximately one every forty-five seconds.
- Each turret fired a full salvo in a specific order: first the left-hand gun, then the right-hand gun, and finally the central one, with a delay of sixty milliseconds. This procedure had an appreciable effect on reducing the effects of dispersion.
- The charge was loaded at 5 degrees and it required just 1.3 seconds for the gun to resume its elevation.
- The charge was composed of 660 pounds of cellulose nitrate and, after 1945, of smokeless powder diphenylamine (SPD) in six bags. The reduced charge consisted of 305 pounds of smokeless powder diphenylamine nonhygroscopic (SPD or SPDN).
- On January 20, 1989, *Iowa* hit a target on Vieques Island at a range of 47,350 yards, the record for a gun of this caliber.
- When firing in company, the nose of each ship's AP shells were filled with colorant to distinguish her fall of shot: *Iowa* = orange, *New Jersey* = blue, *Missouri* = red, and *Wisconsin* = green.

USS *Missouri* in 1940

CHARACTERISTICS AND PROTECTION OF THE MAIN ARMAMENT (16-INCH)

Caliber	50
Elevation	+45° to −5°
Muzzle velocity (AP)	2,425 ft/sec
Theoretical rate of fire	2 shells/min
Barrel life with full charge	290–350 firings, increased to 1,500 with the addition of a wear-reduction jacket placed round each powder bag in the 1980s
Training speed	4°/sec
Elevation speed	12°/sec
Recoil	47 in
Range by elevation	17,750 yds at 10°
	23,900 yds at 15°
	29,000 yds at 20°
	33,300 yds at 25°
	36,700 yds at 30°
	39,500 yds at 35°
	41,430 yds at 40°
	42,345 yds at 45°
Protection	
Face	17 in + 2.5 in—inclined to 36°
Side	9.5 in
Roof	7.25 in
Rear	12 in
Barbette	17.3 in high and 11.6 in above deck 2
	3 in between decks 2 and 3
	1.5 in under deck 3
Crew	77

BALLISTIC CHARACTERISTICS OF THE 16-INCH SHELL

Armor penetration	26 in at 10,000 yds
	20 in at 20,000 yds
	15 in at 30,000 yds
	11 in at 40,000 yds

View of the after superstructure of *Iowa* in 1944.

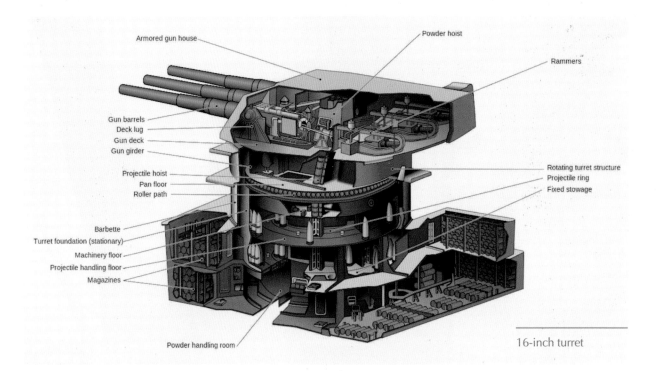

16-inch turret

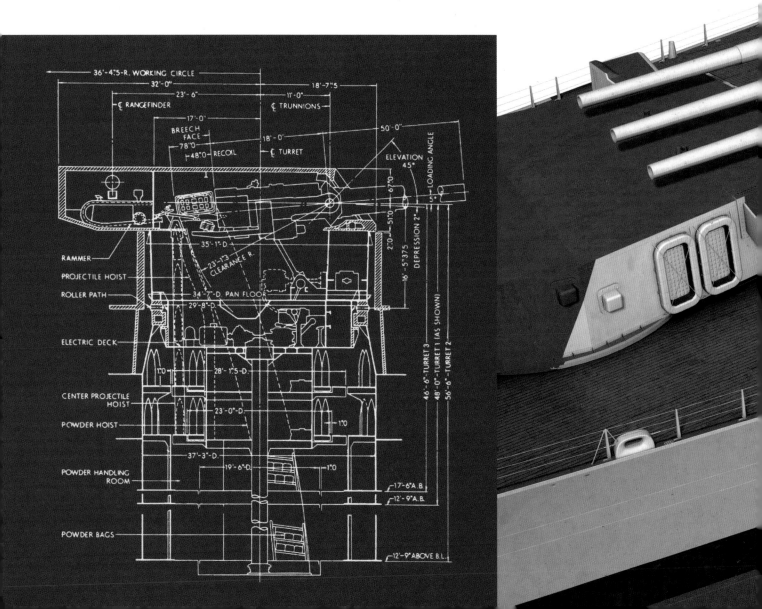

USS *Iowa* in 1944

New Jersey's forward 16-inch turrets.
USN

The upper section of one of *New Jersey*'s turrets being fitted on January 12, 1942.
USN

Components of one of *Wisconsin*'s 16-inch turrets hoisted into position on December 22, 1943.
USN

The rear of *Iowa*'s after 16-inch turret topped by its 40-mm Mk-2 mount on July 9, 1943.
USN

Propellant supply via the powder hatches.
NARA

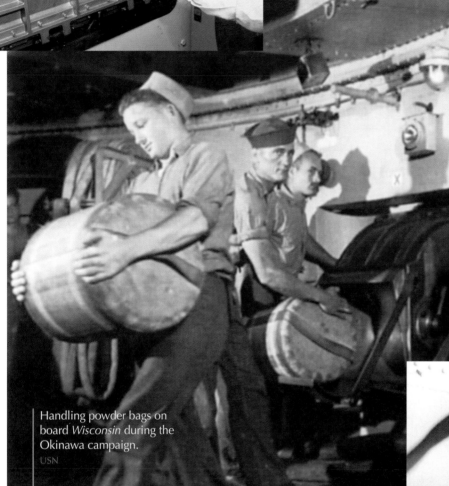
Handling powder bags on board *Wisconsin* during the Okinawa campaign.
USN

A 16-inch shell hoist.
USN

The gun captain inspects the impressive breach of a 16-inch gun in *Wisconsin*, Okinawa, May 1945.
USN

This and the two following pages: The various phases of loading a 16-inch gun.
NARA

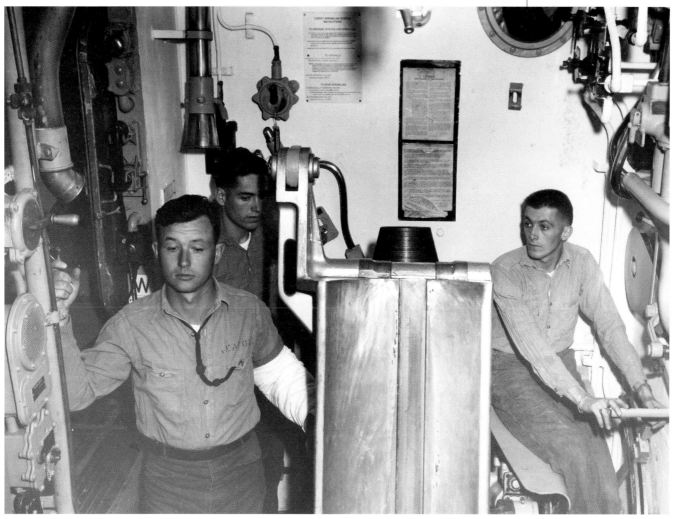

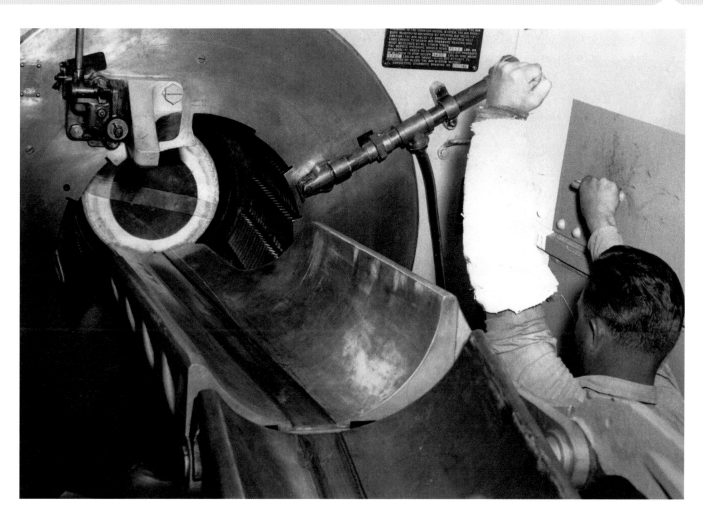

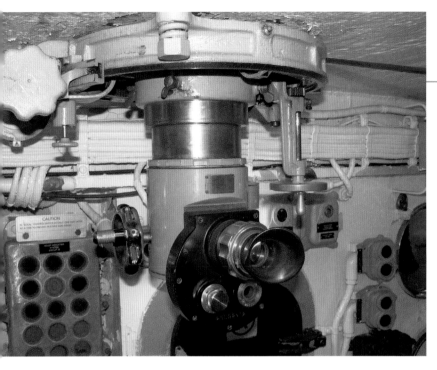

The telescopic sights with which the main turrets were equipped.
PHILIPPE CARESSE

An inflatable target being readied for floating off the stern of the ship prior to gunnery exercises in the 1980s.
USN

Iowa opens up with her main battery.
USN

Iowa fires her 16-inch guns during Baltops 85.
USN

The 5-Inch Secondary Armament

The dual-purpose secondary battery was composed of twenty 5-inch Mk-12 38-caliber guns mounted in Mk-28 Mod 2 mounts weighing 76.9 tons each with 450 shells provided per mount.

The 5-inch shell weighed 55 pounds. Two types of shell were embarked: HE/HC for surface and shore bombardment targets and AAC Common for antiaircraft use.

The cartridges were filled with 15 pounds, 7 ounces of cellulose nitrate and with SPD or SPDN D282 after 1945. The reduced charge consisted of 3 pounds, 8 ounces of SPD or SPDN D282.

CHARACTERISTICS AND PROTECTION OF THE SECONDARY ARMAMENT (5-INCH)

Caliber	38
Elevation	+85° to –15°
AP shell muzzle velocity	2,400 ft/sec
Rate of fire	16 to 23 shells/min
Barrel life with full charge	4,600 firings
Training speed	25°/sec
Elevation speed	15°/sec
Recoil	15 in
Range by elevation	11,663 yds at 15°
	15,919 yds at 30°
	17,392 yds at 45°
AA ceiling	37,200 ft
Protection	
Face and sides	2.5 in
Crew	27

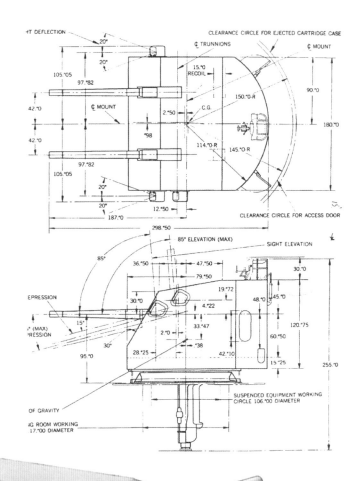

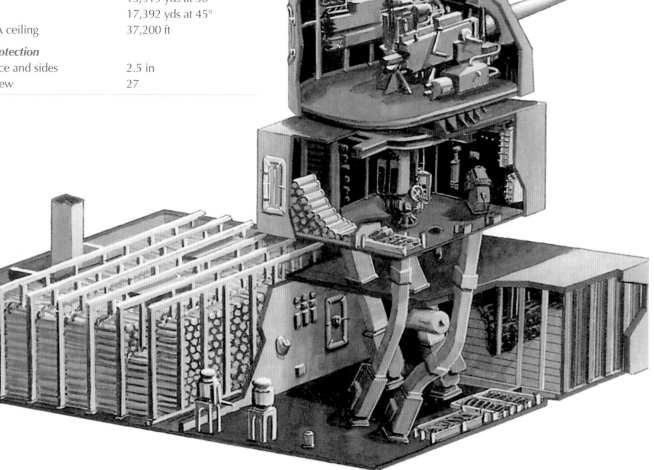

An excellent view of *Iowa*'s starboard 5-inch Mk-28 battery on July 9, 1943.
USN

Maintenance on one of the twin 5-inch mounts.
NARA

Embarking 5-inch shells.
USN

The interior of a 5-inch mount and members of its crew.
NARA

Numerous shell casings bear witness to prolonged firing by the 5-inch guns.
NARA

A 5-inch mount opens fire.
NARA

Iowa's secondary battery opens fire in 1984.
NARA

The 40-mm Antiaircraft Armament

In 1945 antiaircraft defense was provided by twenty (*Iowa*: nineteen) 40-mm Bofors Mk-2 56-caliber guns in quadruple mounts. Together with its protective shield, each mount weighed 11.3 tons, and 1,340 shells were provided per unit.

Three types of shell were embarked: the HE Mk-1, HE Mk-2, and the AP M81A1 or M81A2. The HE shells weighed 1,985 pounds and the AP weighed 1,960 pounds.

Most of these mounts were removed between 1967 and 1968, although some were landed during the Korean War era.

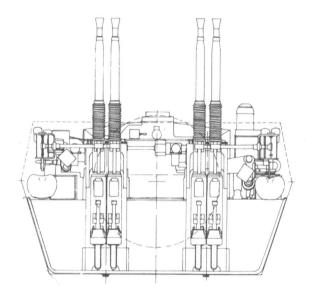

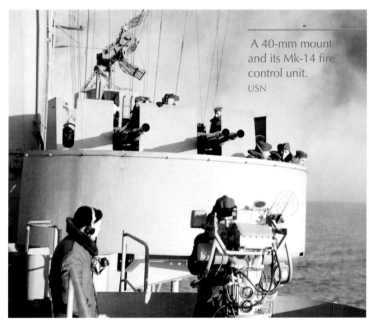

A 40-mm mount and its Mk-14 fire control unit. USN

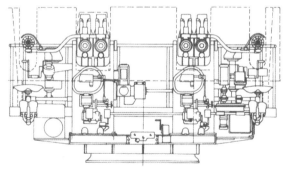

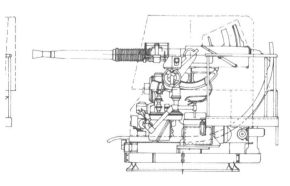

CHARACTERISTICS OF THE 40-MM ARMAMENT

Elevation	+90° to –15°
Muzzle velocity (AP)	2,890 ft/sec
Rate of fire	160 shells/min
Barrel life with full charge	9,500 firings
Training speed	26°/sec
Elevation speed	24°/sec
Recoil	8.7 in
Range by elevation	HE: 8,227 yds at 15° AP: 7,580 yds at 15°
	HE: 10,691 yds at 30° AP: 9,358 yds at 30°
	HE: 11,133 yds at 45° AP: 9,492 yds at 45°
	AA ceiling: 22,300 ft
Crew	11

Three views of the 40-mm Mk-2 Bofors mounts in *Iowa*.
USN

Quadruple 40-mm mounts at close quarters.
USN

New Jersey's 40-mm guns in action in 1953.
USN

The 20-mm Antiaircraft Armament

The close antiaircraft defense consisted of 20-mm 70-caliber Mk-4 mounts. Some 1,500 shells were embarked per unit.

Two types of shell were embarked: HE and AP tracer. The HE and the AP both weighed a shade under 4.3 ounces giving a total weight including the cartridge of 8.5 ounces.

All the mounts were removed in 1952.

CHARACTERISTICS OF THE 20-MM ANTIAIRCRAFT ARMAMENT

Elevation	Mk-4: +87° to −5°
	Mk-10: +90° to −15°
Muzzle velocity	2,740 ft/sec
Rate of fire	250 to 320 shells/min
Barrel life with full charge	9,000 firings
Range by elevation	3,950 yds at 15°
	4,650 yds at 30°
	4,800 yds at 45°
AA ceiling	10,000 ft

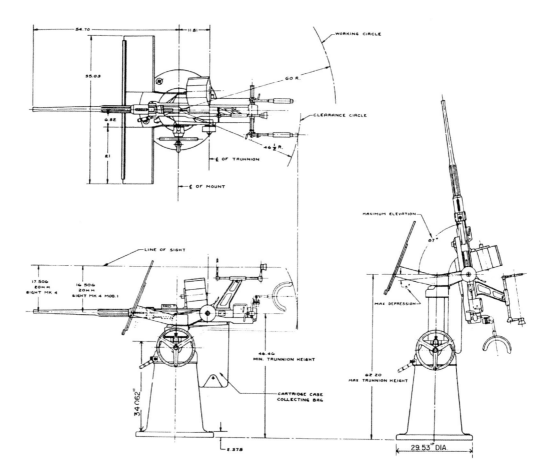

Iowa's 20-mm guns seen on July 9, 1943.
USN

Three of the 20-mm mounts in action with a 40-mm Bofors mount in the foreground.
USN

New Jersey's antiaircraft defense on June 24, 1945: 5-inch mounts, quadruple 40-mm mounts and 20-mm guns.
USN

The Phalanx CIWS

Between 1984 and 1988 the *Iowa* class was equipped with four 76-caliber 20-mm Phalanx Mk-15 Block 0 CIWS (close-in weapon system) mounts. The APDS Mk-149 shell weighed 3.5 ounces. Total weight including the cartridge was 9.3 ounces. The muzzle velocity of the round was 3,650 feet per second. The training and elevation speed was 115 degrees per second, with elevation varying from +80 degrees to –20 degrees. The ceiling at 45 degrees was 18,000 feet. Rate of fire was 3,000–3,500 rounds per minute. There were 8,000 rounds for each unit.

A Phalanx mount on board *New Jersey* (*top*) and *Iowa* (*bottom*).
PHILIPPE CARESSE; USN

Ammunition supply for a Phalanx mount.
USN

A Phalanx in action.
USN

SRBOC launchers and a Phalanx mount in *Iowa* on September 1, 1985. The West German frigate *Bremen* (F 207) is on her starboard quarter.
USN

The Tomahawk System

The BGM-109 Tomahawk cruise missile could be fitted with a conventional or W80 nuclear warhead. It was provided with an inertial guidance and GPS system as well as an infrared seeker head. The Tomahawk had a subsonic cruising speed of 550 miles per hour and a maximum range of 1,500 miles. It had a length of 20 feet 6 inches including the booster and a diameter of 20 inches. The conventional warhead weighed 1,000 pounds.

The Tomahawk system was built by Raytheon of Tucson at a cost of $550,000 per missile.

Three versions were available: the TASM (Tomahawk Anti-Ship Missile), the TLAM-C (Tomahawk Land-Attack Missile Conventional) and the TLAM-N (Tomahawk Land-Attack Missile Nuclear).

Missiles were stored and launched from armored box launchers (ABLs), an aluminum assembly housing four launch tubes that were hydraulically elevated to 35 degrees for firing. Four ABLs were disposed on either beam between the two stacks, with four more positioned abaft the after stack and oriented 45 degrees off the bow.

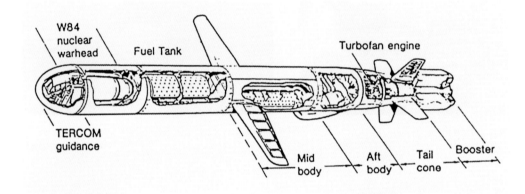

A Tomahawk being inserted into its launcher.
USN

The armored launchers housing Tomahawk cruise missiles.
NATIONAL ARCHIVES; USN

Tomahawk and Harpoon missile launchers seen on September 23, 1985, during *Iowa*'s visit to Cherbourg.
PRIVATE COLLECTION

A Tomahawk fired by *New Jersey* on May 10, 1983.
USN

A Tomahawk firing from *Wisconsin*.
NARA

The Harpoon System

The RGM-84 Harpoon antiship missile was powered by a Teledyne J402 turbojet engine with 2.9kN of thrust assisted at takeoff by a powder accelerator. The guidance system was preprogrammed followed by a self-steering active J band.

The Harpoon had a speed of 528 miles per hour and range of at least 65 miles. It had a length of 12 feet 6 inches, a wingspan of 3 feet, and a diameter of 13.5 inches. The total weight was 1,144 pounds including a 487-pound high-deflagration warhead.

The Harpoon system was built by Boeing Integrated Defense Systems at a cost of $720,000 per unit. Four quadruple Harpoon assemblies were positioned on each side of the after stack, disposed on the beam.

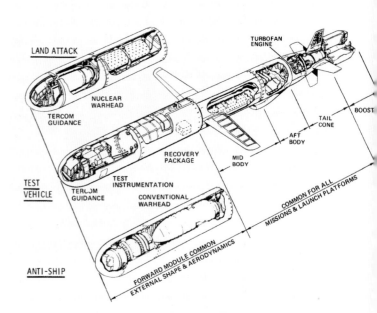

A Harpoon in flight.
NARA

The Harpoon system installed in *Iowa*.
NARA

Armament 79

Harpoon launchers in *Iowa*.
NARA

Self-Defense

From 1983 the *Iowa* class was also equipped with eight sextuple SRBOC Mk-36 (Super Rapid Bloom Offboard Chaff) countermeasures system in the form of a mortar launching decoys at specific heights and ranges against a variety of threats.

SRBOC Mk-36 launchers in *New Jersey*.
PHILIPPE CARESSE

Crewmen awaiting ammunition transfer from the USS *Nitro* (AE 23).
USN

An aerial view of *Missouri*'s superstructure and armament.
J. PRADIGNAC

FOUR

Power and Propulsion

The steam plant was composed of three-part Babcock & Wilcox Type-M boilers superheated to 850 degrees F and rated at 565 psi. They were distributed in four firerooms.

Each turbine with its double reduction gear had its own engine room. *Iowa* and *Missouri* had General Electric reduction gear, while *New Jersey* and *Wisconsin* had Westinghouse equipment.

The engine rooms were located abaft each fireroom, requiring the shafts to be of different lengths: shaft 1 measuring 340 feet; shaft 2 of 243 feet; shaft 3 of 179 feet; and shaft 4 of 277 feet.

The four battleships were always stated as having a speed of 33 knots, but Secretary of the Navy John Lehman confirmed that they could reach a speed of 35 knots. The fastest unit of the class was *New Jersey*, which shortly before her deployment to Vietnam made 35.2 knots, but information on her displacement at that time is lacking. *Iowa* was reportedly still able to reach those speeds in the 1980s with steam in hand.

- By way of example, the fuel consumption of *Missouri* during operations against Okinawa and Kyūshū was as follows: she steamed for 336 hours, covered a distance of 6,014 miles at an average speed of 17.9 knots, and consumed 3,978 tons of fuel oil at an average of 87 tons per hour. During the same period she was replenished by oilers three times during which 5,294 tons of fuel oil was embarked, while distributing 2,045 tons of fuel to seventeen destroyers of her escort.
- The rudders could be moved through an arc of 36.5 degrees on either side and weighed 58.42 tons each.
- The *Iowa* class maneuvered like a destroyer on the high seas but handled more gingerly in the shallows.
- More than 900 electric motors were embarked and power provided by eight 1,250kW turbogenerators together with twelve 250kW diesel generators for a total power of 10,000kW at 450/110V.
- Three desalinating plants were capable of producing up to 120,000 gallons of fresh water per day intended mainly to feed the boilers but also to provide running water for the crew.

THE ENGINE PLANT	
Boilers	8 Babcock & Wilcox
Turbines	4 Westinghouse (New Jersey and Wisconsin)
	4 General Electric (Iowa and Missouri)
Design power	212,000 shp
Power	230,000 shp
Fuel oil capacity	1943: 8,084 t
	1949: 8,841 t
	1968: 7,600 t
Range (1945)	18,000 nm at 12 knots
	15,900 nm at 17 knots
	5,300 nm at 29.6 knots
Diesel capacity (1945)	187.2 t
Speed	33 knots (official)
Propellers	2 outer, four bladed Ø 18 ft 3 in
	2 inner, five bladed Ø 17 ft
Weight	±20 tons
Rudders	2 of 340 sq ft each

Power and Propulsion 83

- The principal and auxiliary engine plant weighed a total of 4,797 tons.
- The main command position was located on deck 3 and commanded the Damage Control Central (DCC), which was responsible for the management of nine safety teams, each of which was composed of twenty men under the command of an officer.
- There were 1,090 telephones, 248 miles of electrical cabling, and 5,300 lighting systems.
- The class was provided with an underway replenishment (UNREP) system for the benefit of its escort, including a gantry installed on the starboard side during the last major refits to facilitate the use of the fuel hoses.
- The class always had a reputation for being very wet ships forward, and during NATO exercises in the Atlantic in August 1953 *Iowa* recorded rolling up to 26 degrees while HMS *Vanguard*, sailing in company, recorded no more than 15 degrees.

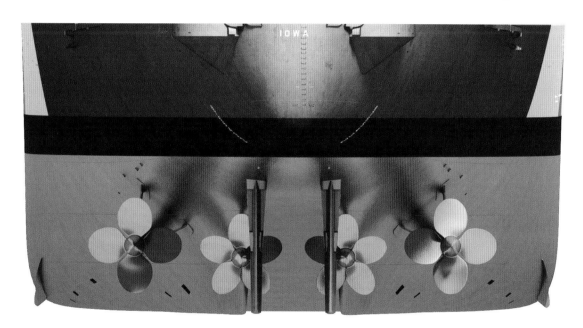

A fireroom in *Iowa* during World War II.
USN

Access to the engine spaces.
USN

The interior of an engine room.
USN

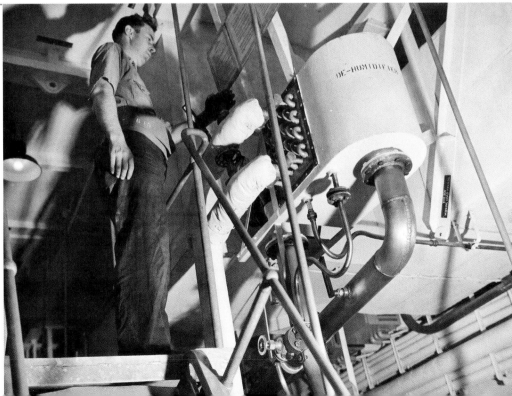

The interior of an engine room.
USN

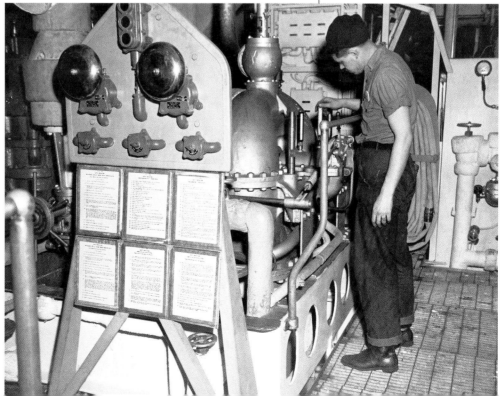

A fireman operating a boiler.
NARA

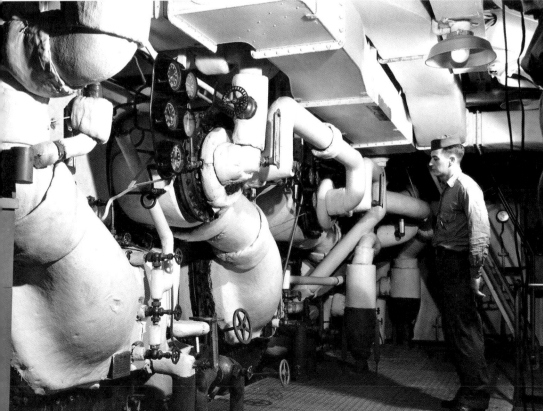

A section of the evaporating equipment.
NARA

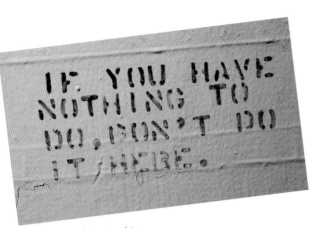

An unambiguous message at the exit to one of the firerooms.
PHILIPPE CARESSE

One of *Missouri*'s engine rooms.
USN

The control panel in one of *New Jersey*'s engine rooms.
NARA

One of the turbogenerators.
NARA

A turbogenerator control panel.
NARA

Recent photographs of a machine shop.
PHILIPPE CARESSE

The gantry and hoses for underway oil replenishment.
USN

Replenishment at sea for the frigate *Trippe* (FF 1075).
USN

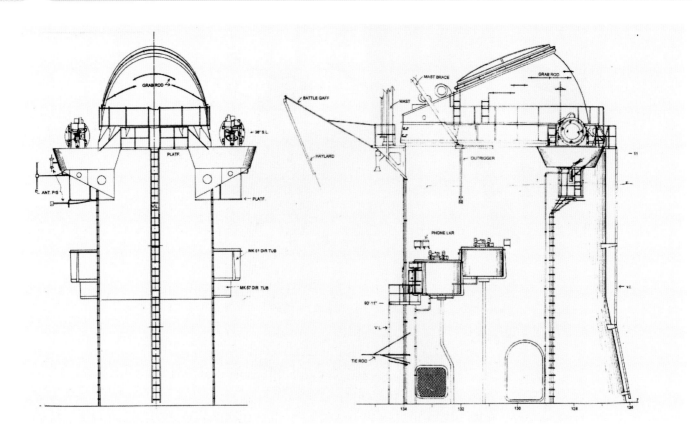

Technical documentation relating to the engine plant of an *Iowa*-class battleship.
USN

Boiler feed water analysis.
NARA

Workers dwarfed by the propellers and rudders of *Iowa* in 1985.
USN

The rudders have a total surface area of 340 square feet.
USN

A 17-foot-diameter propeller. USN

One of *Missouri*'s 18-foot 3-inch-diameter propellers.
USN

The rudders and propellers of *Missouri* seen on July 24, 1944.
USN

New Jersey at sea.
PRADIGNAC AND LEO

FIVE

Shipboard Equipment

The Control Tower and Main Directors

As originally designed, the *Iowa* class had an open bridge leading to a three-level conning tower forward with 17.5 inches of horizontal protection, a 7.25-inch roof, 4-inch deck, and armored tubes for transmitting orders extending to the main command position 16 inches thick. This control position contained a steering position, the command transmitters, ship's log, gyrocompass repeater, and several periscopes. It was accessed by two armored doors and had five viewing slits. The first stage of the conning tower (03 Level) was reserved for the admiral and his staff and the second (04 Level) for the senior officers and their staff, with the pilothouse positioned forward and the chart house aft. The weight of armor was 202.61 tons for the walls (*Iowa* = 300 tons), with the ceiling and deck totaling 33.89 tons and each door weighing 7.05 tons (*Iowa* = 9 tons).

The rounded navigation bridge was enclosed in 1943–44 and large bay windows installed to shelter the men from inclement weather. The bridge was completely redesigned into a rectangular shape in June 1945, with *Missouri* and *Wisconsin* receiving these modifications on commissioning.

The main fire-control position was located on the upper level of the control tower and supported a 26-foot 6-inch rangefinder. It was protected by 1.5 inches of armor to which 1.5 inches was added later.

The secondary fire-control position was located abaft the second stack atop the after main radar position and carried a second 26-foot 6-inch rangefinder.

The forward superstructure of *Iowa* at Boston in November 1943. Note the original bridge and the thickness of the armored door leading to the conning tower.
NARA

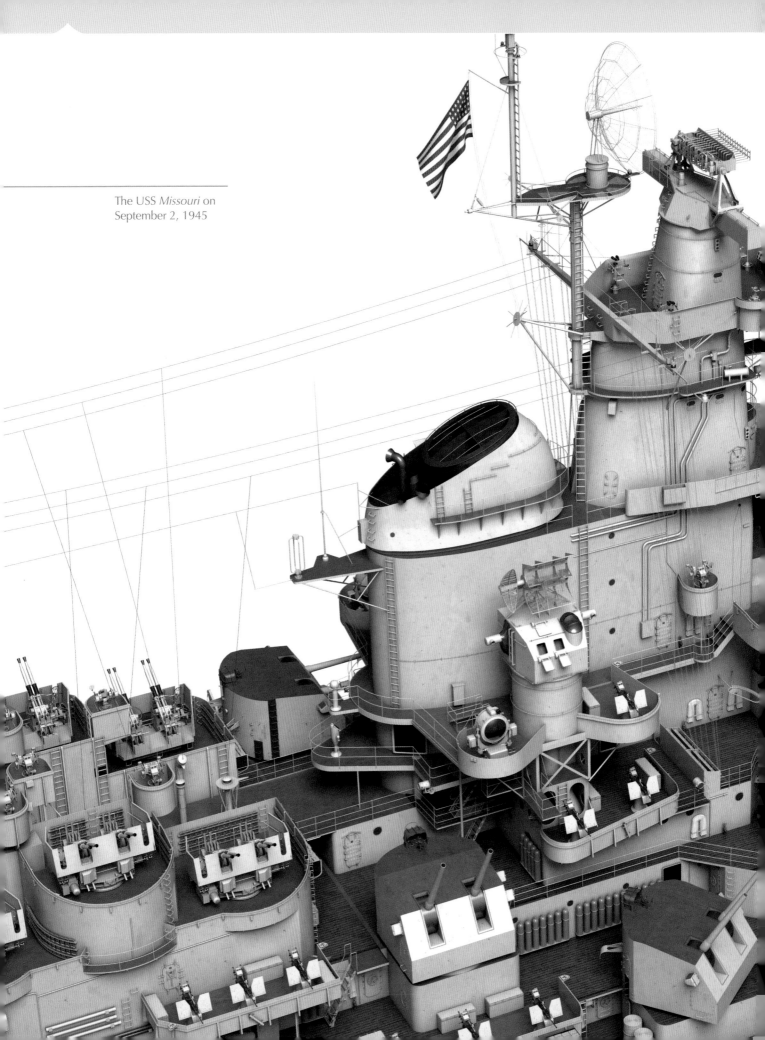

The USS *Missouri* on September 2, 1945

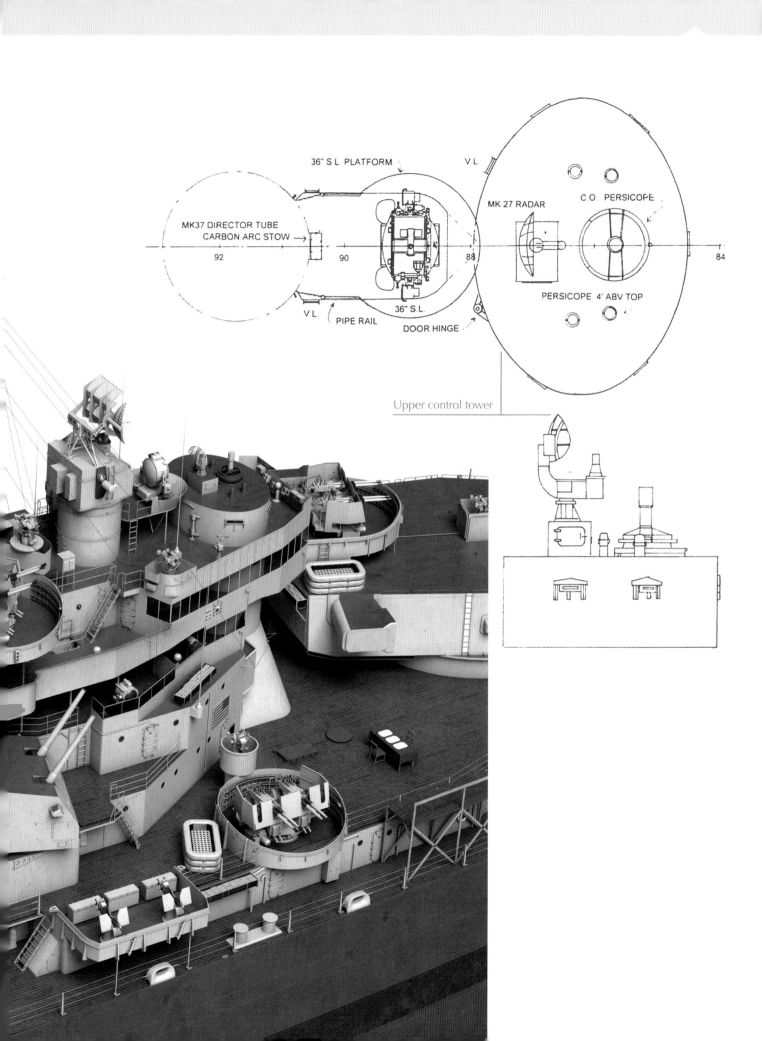

36" S L PLATFORM   V.L.
MK37 DIRECTOR TUBE
CARBON ARC STOW
92   90   88   MK 27 RADAR   C.O. PERSICOPE   84
V.L.   36" S.L.   PERSICOPE 4' ABV TOP
PIPE RAIL   DOOR HINGE

Upper control tower

The forward superstructure of *New Jersey* on June 24, 1945. The bridge is in its final configuration.
NARA

The rounded bridge of *New Jersey* on October 11, 1943.
USN

A bird's-eye view of the conning tower, the bridge, and its various installations in 1945.
USN

Missouri's bridge seen in August 1944.
USN

A bird's-eye view of the conning tower, the bridge, and its various installations in 1945.
USN

Missouri's bridge seen in August 1944.
USN

The passage running along the rear of the conning tower.
PHILIPPE CARESSE

The shaft revolution indicator on the open bridge.
PHILIPPE CARESSE

The interior of *Missouri*'s bridge in 1944.
NARA

The interior of *Missouri*'s bridge in 1984.
NARA

Inside the conning tower during the Pacific campaign.
NARA

A passage abaft the bridge with a viewing slit set into the armor protecting the 03 Level Signal Shelter on the left.
USN

Shooting the sun on *Missouri*'s bridge.
USN

A navigating officer in the chart room.
NARA

The compass binnacle abaft the forward stack.
PHILIPPE CARESSE

Meteorologists at work on the fantail in January 1985.
USN

The flag deck.
NARA

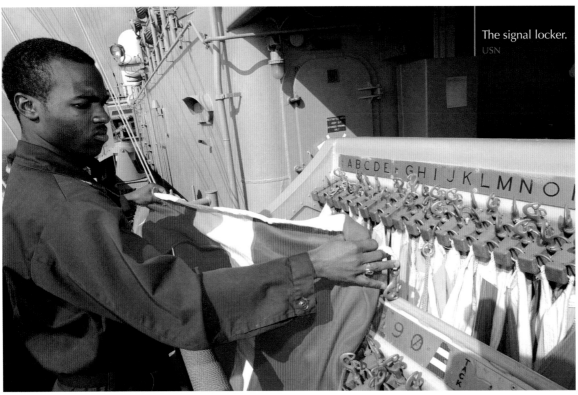

The signal locker.
USN

The flag deck.
NARA

Access to the bridge from the 03 Level of the conning tower.
NARA

New Jersey at anchor on February 8, 1945. Mounted on the conning tower is the Mk-3 radar aerial. Above the Mk-37 directors are the Mk-4 aerials, with the SK aerial visible atop the foremast and the SG surface-search radar aerial atop the mainmast.
USN

Radars and Electronics

- The units of the *Iowa* class were fitted with gunnery and navigation radar from the time of commissioning.
- Over the course of their careers the class saw the installation of SK, SK-2, SR, SR-3, SPS-6, SPS-12, and SPS-49 air-search radar units with a range of between 120 and 250 miles.
- The high-altitude search radar was the SP and SPS-8.
- Surface-search radar included SG, SG-6, SQ, SU, SPS-4, SPS-10, SPS-53 (only in *New Jersey* in 1968) and SPS-67.
- The navigation radar was the SPS-10.
- Fire control for the 16-inch guns was assisted by Mk-8 radars with a range of 40,026 yards and by Mk-13 units between 1945 and 1952.
- In 1943 *Iowa* and *New Jersey* were equipped with Mk-4 radars for the secondary armament with a range of 40,463 yards. In 1945 the four battleships had Mk-12/22 radar with a range of 44,838 yards, and the Mk-25 unit from 1948–55.
- The 40-mm antiaircraft armament operated with Mk-19 and Mk-29 radar (1945) and then Mk-35 radar (1954–55).
- The CIWS operated with VPS-2 Pulse Doppler radar mounted on the turret and rotated with the guns. It had a range of 5,468 yards.
- The Tomahawk was programmed before launching with heading and course alterations to allow for any changes in the launch position. Target guidance was provided by the TERCOM (TERrain COntour Matching) inertial navigation system combining an inertial unit or GPS receiver with an altimetric radar, allowing the missile to follow contours with precision. For surgical strikes a system such as the DSMAC (Digital Scene Matching Area Correlator) was utilized for final guidance to land targets.

- The Harpoon guidance system was composed of an active searcher, a radar and its radome, a missile guidance unit (MGU), a radar altimeter, and a power converter. The MGU consisted of an altitude reference assembly (ARA) linked to a digital computer power supply (DC/PS). The DC/PS was initialized by the launch system before firing. After launch the DC/PS utilized acceleration data from the ARA and altitude information from the radar altimeter to keep the missile on the programmed flight path. Once the target had been acquired the DC/PS guided the missile to its objective.

- Communications were provided by the NTDS system composed of two AN/SRA-57 and AN/SRA-58 aerials. Satellite communications (Satcom) were provided by two AN/WCS-3 transmitters/receivers and OE-82C/WSC-1 aerials.

RADIO CALL SIGNS:

Iowa: November—Echo—Papa—Mike
New Jersey: November—Echo—Papa—Papa
Missouri: November—Charlie—Bravo—Lima
Wisconsin: November—Uniform—Golf—Whiskey

The foremast of *Iowa* in January 1952. Occupying the highest position is the SPS-6 radar antenna with the SPS-12 antenna just below it. Forward of these is the Mk-13 radar aerial atop the main rangefinder.
NARA

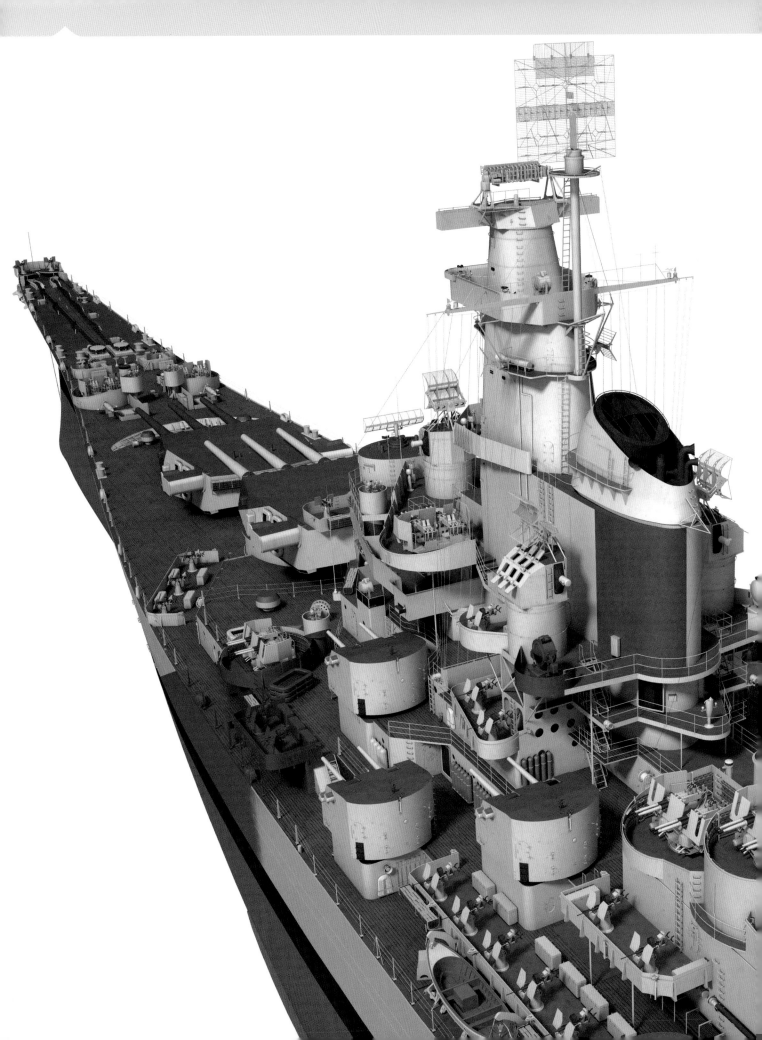

Missouri's SK-2 radar aerial.
USN

This photo, taken in the 1980s, shows the SLQ-32 (V)3 ECM equipment positioned on either side of the main rangefinder together with the OE-8L (Satcom) aerial in the center.
USN

The after stack with the OE-8L (Satcom) aerial.
USN

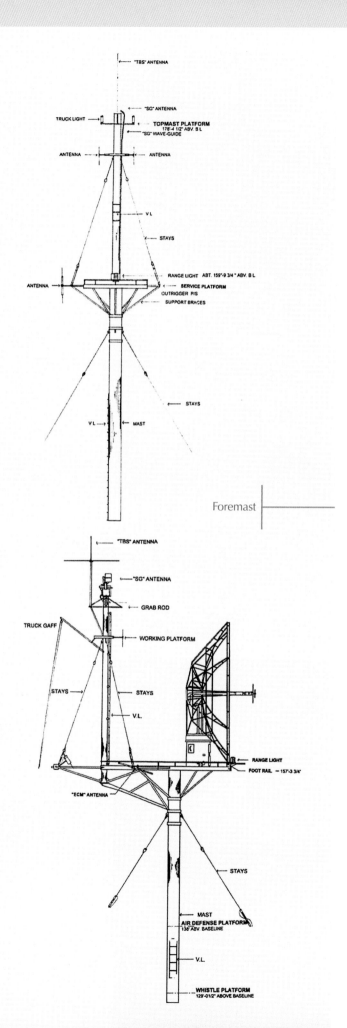

Foremast

The aerials on the after stack.
NARA

Iowa's discone/discage antenna mounted in the bow. Note the ship's bell suspended from it.
NARA

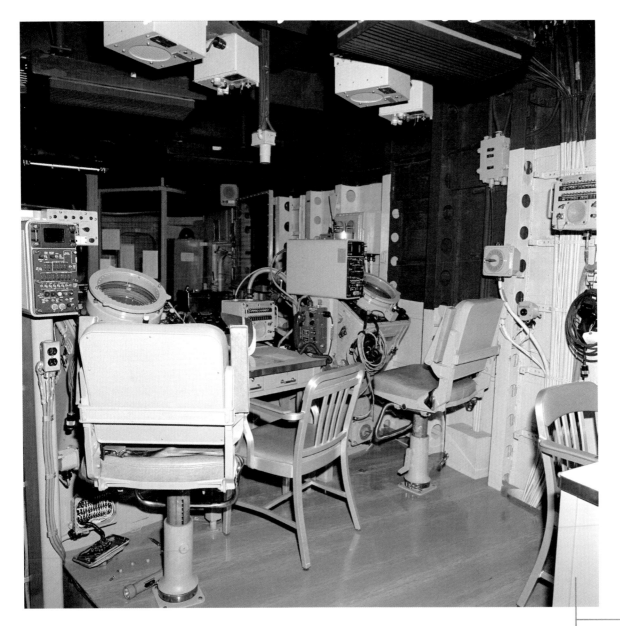

The radar room with its many screens in the 1950s (*top*) and in 1984 (*bottom*).
NARA

Shipboard Equipment   123

Rangefinders

The main rangefinders for the 16-inch guns were in the Mk-38 director units mounted in the superstructure forward (known as spot 1) and aft (spot 2). They were responsible for ranging and spotting the fall of shot, with the information being sent to the two battery plot/fire-control rooms. The Mk-38 directors also contained various sighting telescopes. Backup units were mounted in the turrets, consisting of 46-foot Mk-52 coincidence rangefinders in turrets 2 and 3 with an Mk-53 unit of the same size in turret 1, although the latter was removed from all vessels between 1948 and 1952 as a result of leakage in heavy seas. Each turret was equipped with four Mk-66 telescopes visible on either side, one for the trainer and the other for the pointer, two per side.

The 26-foot 6-inch units fitted atop the control tower and in the after superstructure sat 116 feet and 68 feet above the waterline, respectively; stereoscopic Mk-48 GFCS Mods 1–5 were fitted in *Iowa* and *New Jersey*, and Mk-38 Mods 6–7 in *Missouri* and *Wisconsin*.

Fire control for the 5-inch battery was provided by four Mk-37 units on 05 Level, one forward, one aft, and one on either side. The 5-inch mounts were fitted with four 15-foot Mk-42 rangefinders. Three types of radar were used with the Mk-37 at different times: the Mk-4, Mk-12/22, and the four Mk-25 dishes still mounted on *Iowa*'s Mk-37s and known as Sky 1, 2, 3, and 4.

The 40-mm Bofors guns were supported by Mk-51, Mk-49, Mk-57, Mk-56, and Mk-63 units.

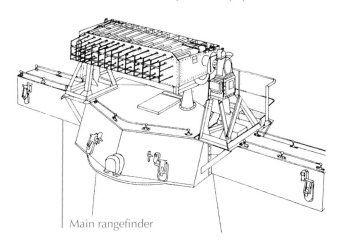
Main rangefinder

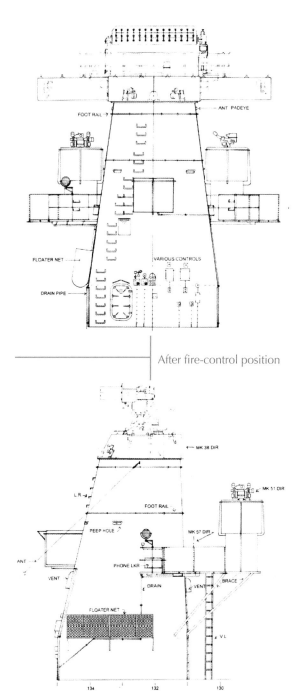
After fire-control position

Iowa's main rangefinder seen atop the foremast in January 1952.

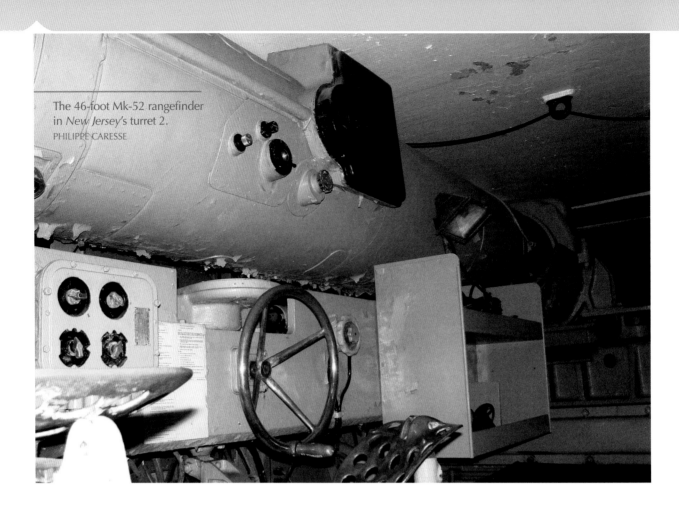

The 46-foot Mk-52 rangefinder in *New Jersey*'s turret 2.
PHILIPPE CARESSE

The optics of a turret rangefinder.
USN

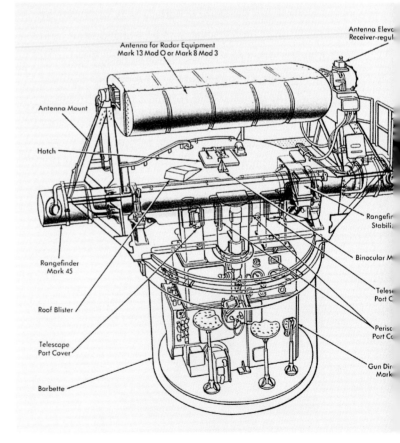

The Mk-45 rangefinder

A Mk-37 rangefinder and its Mk-12 radar aerial in *Missouri* on July 23, 1944.
NARA

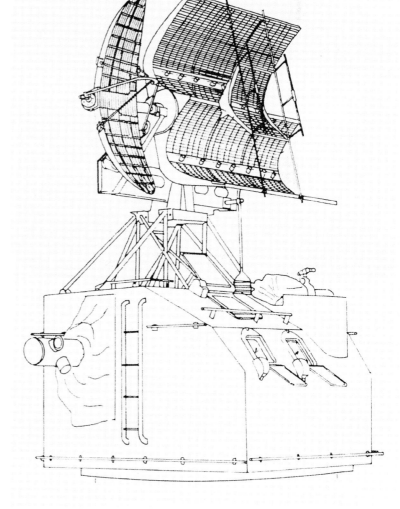

The Mk-37 rangefinder

Iowa's forward Mk-37 rangefinder and Mk-25 radar aerial seen in August 1984.
NARA

A Mk-37 rangefinder and its Mk-25 aerial.
NARA

A view of *Iowa*'s after Mk-37 Sky 4 5-inch director and its crew on August 15, 1984. The unit has been trained on a forward bearing.
NARA

Iowa in the 1950s.
USN

The technical records office.
USN

A secondary battery plotting room presents a hive of activity.
NARA

Information provided by the rangefinders was transmitted to two plotting rooms located on platform 1 forward and on deck 3 aft, both protected within the citadel. These spaces housed an Mk-8 electromechanical rangekeeper that gave a continuous assessment of the training and elevation of the guns and the evolution of the target, incorporating data provided by the gunnery officer as well as by the radar. It also contained a stable vertical Mk-41 gyroscope offering a continuous readout of the angles between the vertical and horizontal planes. This information was transmitted to the rangekeeper in order for the firing solution to be calculated. Shore bombardment against static targets was calculated using an Mk-48 calculator, while an Mk-13 FC radar assessed fall of shot to correct fire. A parallax corrector allowed data to be adjusted between the turrets and the main director. The outfit was completed by a telephone exchange and equipment for the second gunnery officer.

Plotting room

The Mk-8 electromechanical calculator

One of *Missouri*'s two plotting rooms seen on September 17, 1950.
NARA

The Operations Center in 1984.
NARA

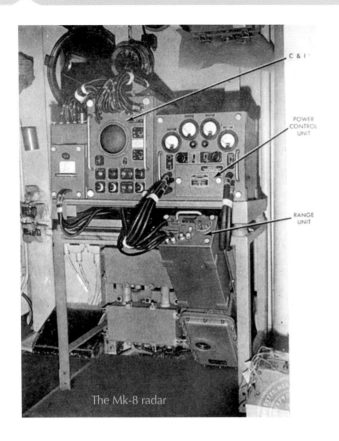

The Mk-8 radar

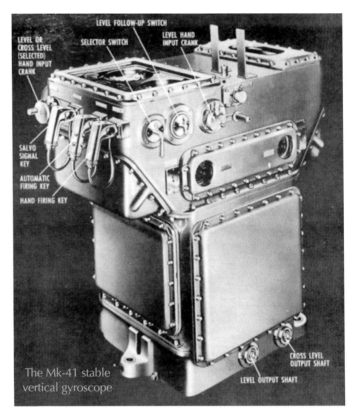

The Mk-41 stable vertical gyroscope

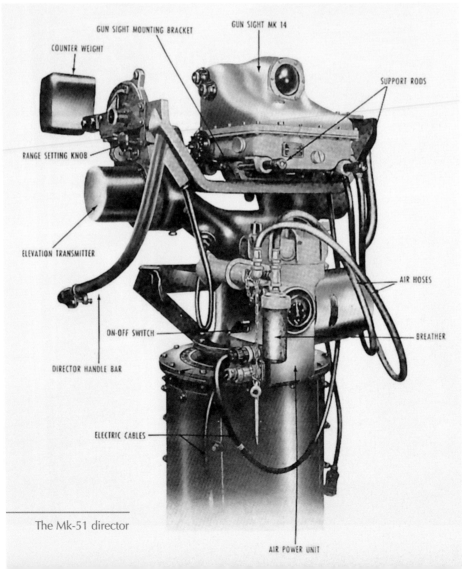

The Mk-51 director

Embarked Aviation

From 1943 to 1945 the *Iowa* class embarked three Vought OS2U Kingfisher reconnaissance and antisubmarine floatplanes crewed by a pilot and radio operator/gunner. Between 1945 and 1947 they were replaced by two single-seat Curtiss SC-1 Seahawk reconnaissance floatplanes in each ship.

The floatplanes were launched by two Type-P Mk-6 catapults positioned on each side of the fantail. The aircraft were handled by a crane with a lifting capacity of 4.21 tons. There was no hangar and the cranes were removed from all units during the 1980s.

From 1951 to 1953 the fantail of each unit in the class was modified to receive the Sikorsky HO3S-1 Horse to which the Piasecki HUP-2 Retriever helicopter was added. There are also photographic records of the Bell H-13 helicopter on *Iowa*'s fantail.

During the 1980s provision was made for a total of seven different types of helicopter to land on the fantails of the units of the class, including the Kaman SH-2 Seasprite, Bell UH-1 Iroquois Huey, Boeing Vertol CH-46 Sea Knight, Sikorsky CH-53 Sea Stallion, Sikorsky CH-53E Super Stallion, Sikorsky SH-60 LAMPS MK III Seahawk, and the Sikorsky SH-3A Sea King. These aircraft could be refueled and minor repairs carried out.

In December 1986 the RQ-2B Pioneer remotely piloted vehicle (RPV) was first deployed in *Iowa* to develop procedures for the other battleships. Launched on a rocket-assisted catapult as a tactical drone and recovered in a net on the helicopter fantail, it was used for observation with a camera and also had a laser designator to assist other platforms with designating targets.

RQ-2B PIONEER

Powerplant	Sachs & Fichtel SF2–350
Power	26 hp
Maximum speed	110 mph
Ceiling	15,000 ft
Weight	452 lb
Wingspan	16.9 ft
Length	14 ft
Height	3.9 ft
Endurance	5 hrs

New Jersey's two Mk-6 catapults and the 4.2-ton handling crane seen on June 24, 1945.
NARA

Catapult

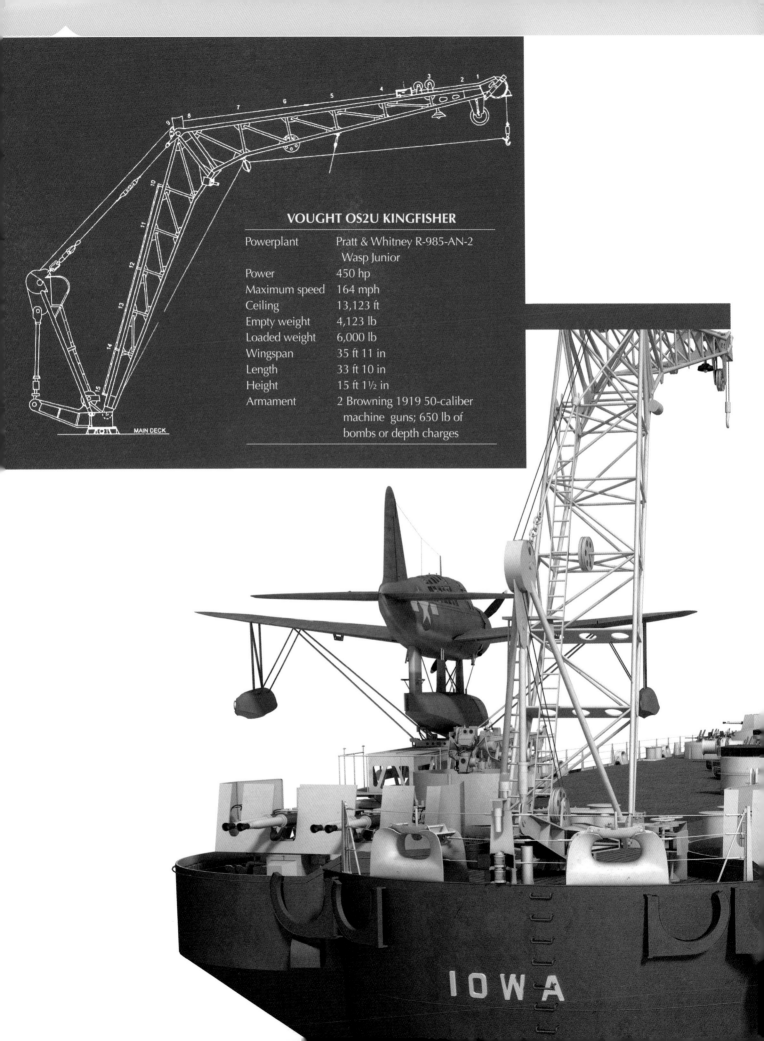

VOUGHT OS2U KINGFISHER

Powerplant	Pratt & Whitney R-985-AN-2 Wasp Junior
Power	450 hp
Maximum speed	164 mph
Ceiling	13,123 ft
Empty weight	4,123 lb
Loaded weight	6,000 lb
Wingspan	35 ft 11 in
Length	33 ft 10 in
Height	15 ft 1½ in
Armament	2 Browning 1919 50-caliber machine guns; 650 lb of bombs or depth charges

Catapult launch of an OS2U Kingfisher floatplane from Missouri.
NARA

Kingfisher pilot prepares for launch.
NARA

CURTISS SC-1 SEAHAWK

Powerplant	Wright R-1820–62 Cyclone
Power	1,350 hp
Maximum speed	313 mph at 28,600 ft
Ceiling	37,303 ft
Empty weight	6,320 lb
Loaded weight	9,000 lb
Wingspan	41 ft
Length	36 ft 4½ in
Height	16 ft
Armament	2 × Browning M2 50-caliber machine guns; two 325-lb bombs

A Curtiss SC-1 Seahawk shortly after launch. Sadly, no examples of this aircraft have been preserved for display.
USN

A helicopter landing pad has replaced the catapults and handling crane on *Iowa*'s fantail.
USN

A Sikorsky HO3S-1 approaching (*top*) and parked (*bottom*) on the fantail of *Iowa* in 1952.
NARA

SIKORSKY HO3S-1

Powerplant	Pratt & Whitney R-985
Power	450 hp
Maximum speed	106 mph
Ceiling	9,842 ft
Loaded weight	4,825 lb
Rotor diameter	48 ft
Length	57 ft 1 in
Height	13 ft

A Kaman SH-2 Seasprite on board *Iowa* on August 1, 1986.
NARA

KAMAN SH-2 SEASPRITE

Powerplant	2 General Electric T58-GE-8F
Power	2 x 1,350 hp
Maximum speed	162 mph
Ceiling	22,506 ft
Maximum takeoff weight	10,200 lb
Rotor diameter	44 ft
Length	52 ft 2 in
Height	13 ft 6 in
Armament	2 antiship missiles or 2 torpedoes

A preflight briefing.
USN

A Sea Knight delivering supplies on *Iowa*'s fantail on July 1, 1984.
NARA

A CH-46 Sea Knight on the fantail of the same battleship.
USN

Refueling an SH-60B Sea Hawk.
USN

A Sikorsky CH-53 Sea Stallion briefly lands on *Iowa* on August 1, 1985.
USN

Aviation technicians at work.
NARA

An RQ-2B Pioneer drone on *Iowa*'s fantail in 1986.
NARA

Launch of an RQ-2B Pioneer drone.
USN

Controlling a Pioneer drone.
USN

Recovery of a drone on the fantail.
USN

Searchlights

Between 1943 and 1945 a total of six 36-inch searchlights were installed, as follows: The first was positioned on the upper bridge between the conning tower and the base of the Mk-37 director in 1943. In 1945 a second unit was installed between the Mk-37 and the bridge, with four more positioned on platforms on either side of each funnel. All were removed between 1951 and 1953.

A 36-inch searchlight in *Missouri* in August 1944.
NARA

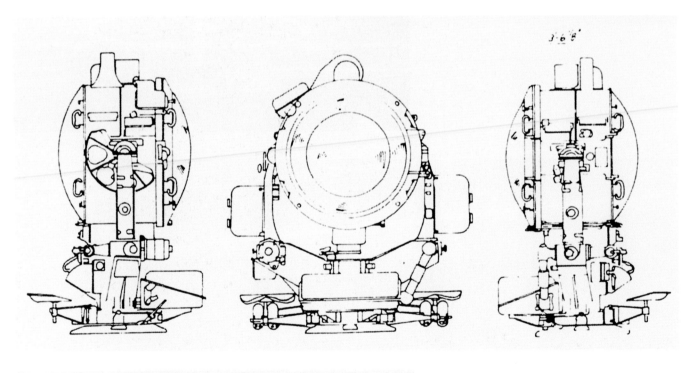

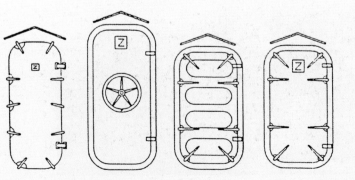

Watertight doors

Exchange of signals by Scott lamp.
USN

The forecastle and
its mooring lines

Ground Tackle

The ships of the *Iowa* class were equipped with two stockless bower anchors weighing 30,000 pounds each. The anchor chains were three inches in diameter with each link weighing 128 pounds. There were twelve 90-foot shots or lengths of chain per anchor line totaling 1,080 feet or 187 fathoms. The chains were run out of chain lockers through hawse pipes on either side of the bow and through the large castings set in the hull known as bolsters or bolster plates.

There were two windlasses on the forecastle, each with a pull weight of 31,500 pounds and powered by 225-hp General Electric electro-hydraulic engines. The capstans were used mainly for warping (i.e., hauling the ship's berthing lines in to tie up to a wharf) and hauling in towing hawsers. They were capable of exerting a pull of 15,000 pounds at a rate of 50 feet per minute in the case of a ten-inch circumference manila hawser or 130 feet a minute for a light line.

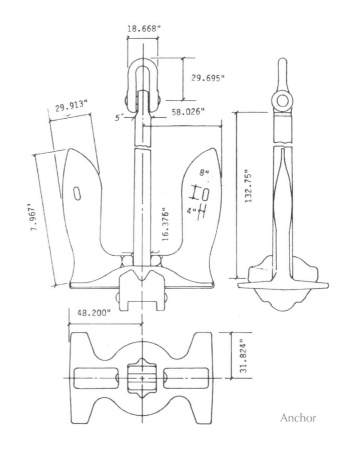

Anchor

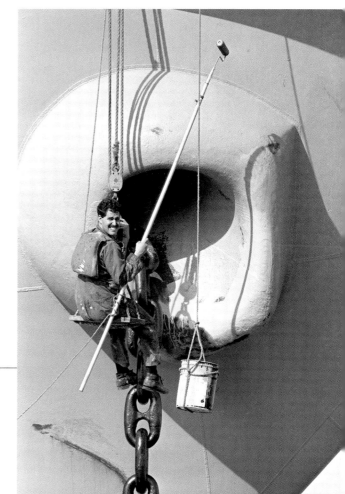

Painting the hawsehole (*right*) and anchor (*above*) on November 1, 1985.
USN

Missouri's port 13.6-ton anchor in 1944.
NARA

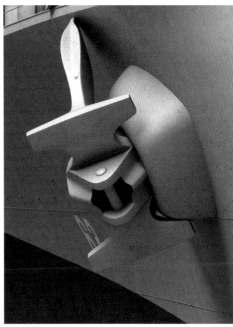

Missouri's 3.5-inch anchor chains.
NARA

Missouri's capstans.
NARA

The capstan

The forecastle with its various equipment.
USN

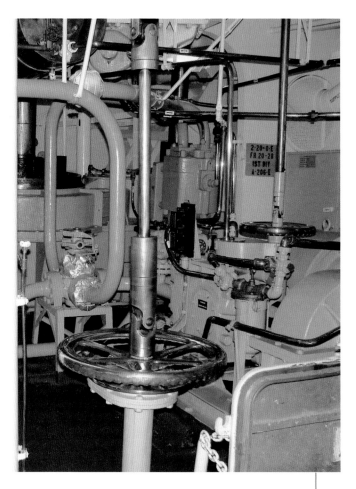

The anchor windlass and capstan engines in 2008.
PHILIPPE CARESSE

Ship's Boats

On commissioning, the *Iowa* class was equipped with two 33-foot Mk-2 boats on davits at frame 114–121 on either side of the main deck abreast the 5-inch mounts. Eighty-one Mk-6 rafts were also positioned on the sides of the 16-inch turret and against the superstructure. After World War II each ship embarked three Mk-3 40-foot launches and two Mk-10 26-foot boats.

In the case of *Iowa* herself, the ship had the following outfit of boats during World War II:

- Two 26-foot motor whaleboats stowed port and starboard on the main deck amidships at frames 114–121. They could be stowed suspended from the boat davits in either an inboard position over the deck or an outboard position over the side. Hoisting and lowering was carried out on lines from the deck winches on the main deck at frames 111–112, port and starboard.
- Two 50-foot motor launches stowed port and starboard on keel brackets on the main deck aft at frames 138–151.
- Two 40-foot motor launches stowed in the 50-foot motor launches.
- Two 35-foot motorboats stowed port and starboard on dollies on the main deck aft at frames 172–182.
- Two 28-foot personnel boats stowed port and starboard on dollies on the main deck aft at frames 145–155 (or 172–181 starboard and 181–189 port).
- The boats stowed on the main deck aft were all handled by the deck winches on the 02 Level at frames 131–139, port and starboard.
- Fifty-nine Carley floats (rafts) and forty-three buoyant life nets with a capacity of twenty-five were stowed throughout the weather decks. These brought the total capacity of the ship's boats to 2,550.

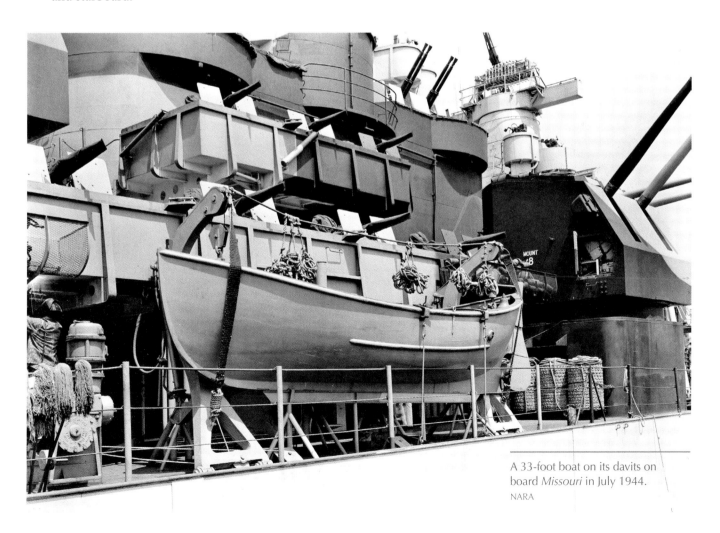

A 33-foot boat on its davits on board *Missouri* in July 1944.
NARA

Based on photographs of *Iowa* during the Korean War, two motor whaleboats were still being carried amidships, port and starboard. No boats were stowed in the vicinity of the after superstructure, but at least six boats of different types were stowed on dollies on the fantail abaft turret 3.

Various boats were carried over the years to transport the crew to and from the ship when she was anchored offshore. The port boat cradle installed in the 1980s normally held a 40-foot Mk-3 utility boat (UB) with a capacity of seventy-five, together with a 26-foot Mk-10 whaleboat stowed above it. Another cradle on the starboard side held a 33-foot Mk-2 officer's motor barge (OMB) with a capacity of twenty-four, together with a 26-foot whaleboat stowed above it with a capacity of twenty. Two 40-foot Mk-3 utility boats and a 33-foot Mk-2 OMB were stowed on dollies on the main deck near the after port superstructure to serve as the captain's gig. If an admiral was embarked, the starboard OMB would become the admiral's barge.

The after port boat boom fitted on the superstructure had a lifting capacity of 10 tons and was used for hoisting motor launches to and from the main deck aft. The hoisting winch used to operate the boat boom was forward of turret 3, near the port superstructure.

Inflatable emergency life rafts for the crew were distributed on storage racks throughout the ship. There were approximately eighty-one inflatable Mk-6 life rafts, each with a capacity of twenty-five. The total capacity of the ship's boats and lifeboats was 2,025.

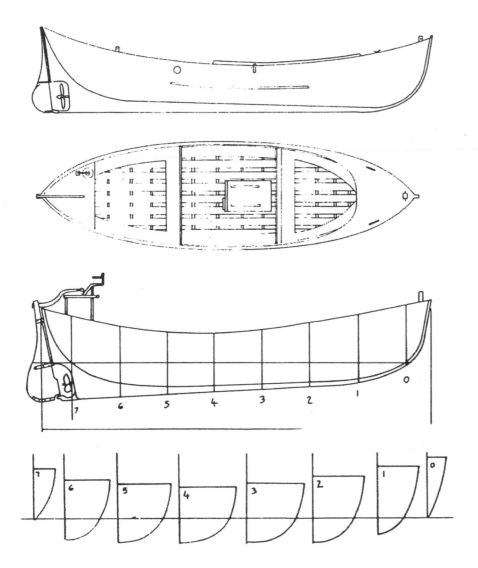

The Mk-2 boat

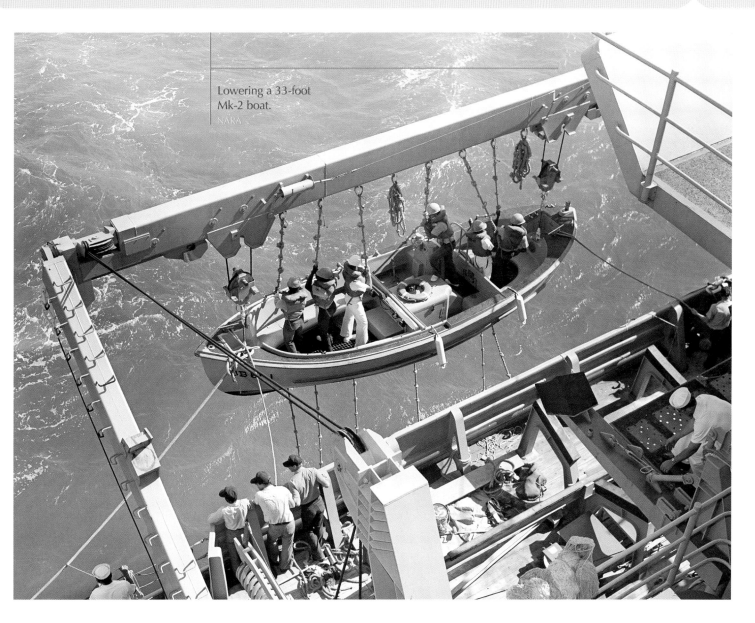

Lowering a 33-foot Mk-2 boat.
NARA

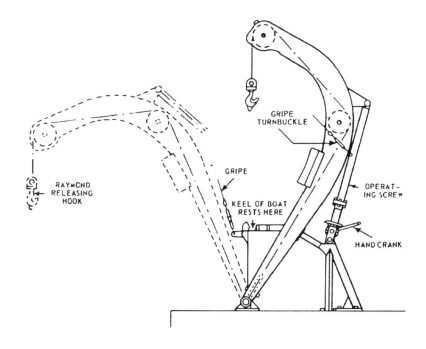

A 40-foot Mk-3 launch.
NARA

Mk-6 rafts on *Missouri*.
NARA

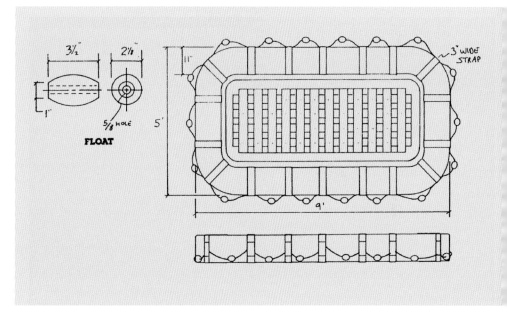

Numerous vessels were stowed behind turret 3 from the 1950s.
J. PRADIGNAC

Provisions

The cold stores could hold 100 tons of fresh fruits and vegetables, 650 tons of nonperishable foodstuffs, and 84 tons of frozen meat. To this must be added the 776 tons of drinking water and 490 tons of running water available for the crew.

Six weeks of provisions were stowed in time of war but the ship could be provisioned for up to sixteen weeks.

In 1945 more than seven tons of food was prepared each day at a cost of $1,600, equivalent to $22,132 in 2018.

The bakery.
NARA

The enlisted men's galley of *Iowa* in 1986.
NARA

A cold store in 1986.
USN

Stowage of
bags of flour.
USN

Fresh milk being brought
on board *Missouri*.
USN

The enlisted men's canteen.
NARA

The enlisted men's canteen in *New Jersey* in 1968.
U.S. NAVAL INSTITUTE PHOTO ARCHIVE

An enlisted men's canteen in *Iowa* in 1956.
NARA

A recent photo of the canteen in *Iowa*.
PHILIPPE CARESSE

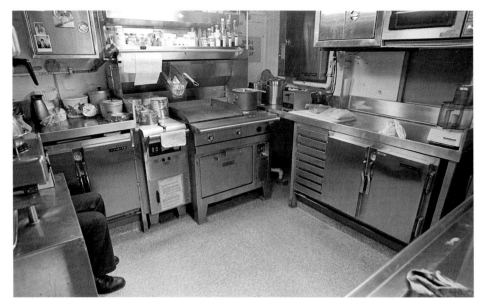

The officers' wardroom galley in the 1980s.
USN

The officers' wardroom in the 1980s.
USN

Complement

The design complement of each vessel was 117 officers and 1,804 enlisted men while the actual complement was as follows:

		IOWA	NEW JERSEY	MISSOURI	WISCONSIN
1945	Officers	151	161	189	173
	Enlisted men	2,637	2,592	2,789	2,738
1949	Officers	166	234	151	169
	Enlisted men	2,451	2,554	2,255	2,503
1968	Officers	—	70	—	—
	Enlisted men		1,556		
1988	Officers	65	65	65	65
	Enlisted men	1,445	1,453	1,450	1,450

A majority of the crew was accommodated on decks 2 and 3. Due to the significant amount of equipment embarked during the last two years of World War II, particularly antiaircraft armament, the accommodation spaces were overcrowded with sparse comfort by the standards of vessels of the latest generation, although each sailor had his own berth and no hammocks were embarked. Officers also suffered comparatively cramped accommodation in their cabins. The living spaces were, however, provided with air conditioning from the late 1960s in the case of *New Jersey* and from 1982 in her sisters, and the class was equipped with the CHT (collection, holding, and transfer) sewage system in 1982.

During the World War II and Korean War eras, more than 3,600 uniforms passed through the hands of the laundrymen each month while the two barbershops on board gave an average of 7,400 haircuts during the same period. Meanwhile, the cobblers replaced 650 pairs of heels and 250 pairs of soles per month.

The captain's in-port cabin in the late 1940s (*left*) and the 1980s (*right*). Captains had two sea cabins on board, one on the 04 Level bridge and the other on the 08 Level battle bridge.
NARA

The officers' wardroom during World War II.
PRIVATE COLLECTION

An officers' head (bathroom).
PHILIPPE CARESSE

The bathtub specially installed for President Roosevelt in *Iowa*.
PHILIPPE CARESSE

An officer's stateroom.
NARA

A coffee break in a petty officers' mess in 1944.
NARA

A petty officers' berthing compartment.
USN

Enlisted men's quarters in *Iowa* in 1986.
NARA

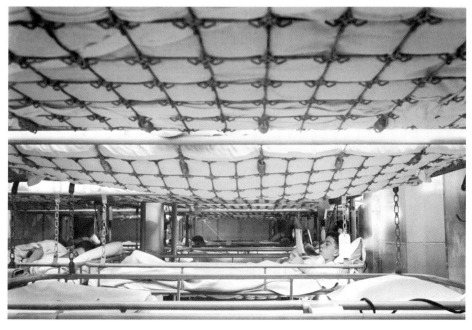

Enlisted men's quarters in *Wisconsin*.
USN

A sailor's locker.
USN

A water fountain in one of the crew quarters of *Missouri*.
NARA

Bunks stowed in one of *Wisconsin*'s administrative spaces.
USN

The "geedunk."
NARA

The laundry in 1986.
NARA

The laundry.
NARA

The brig.
PHILIPPE CARESSE

The administrative offices of *Iowa* in 1987.
NARA

One of *Iowa*'s spare parts stores in 1986.
NARA

The "minimarket" or ship's store in *Iowa*.
NARA

The ship's band.
NARA

Calisthenics to musical accompaniment on the fantail on October 26, 1944.
USN

A basketball game in progress on the starboard side.
USN

Taking receipt of the mail.
USN

The media and audiovisual room for crew entertainment purposes.
USN

The shipboard library of *Wisconsin*.
USN

Crossing the line in *Missouri* in 1986.
USN

Painting the stars and stripes on the roof of turret 1.
USN

Scrubbing *Iowa*'s fantail.
USN

Splicing a heavy hawser.
USN

A snowman takes shape on *Iowa*'s fantail.
USN

A religious service in *Iowa*.
NARA

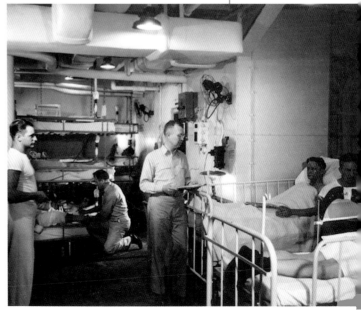

The infirmary of *Missouri* in 1944.
NARA

The Medical Department

Intended to serve as Task Group flagships, the *Iowa* class was equipped with the latest medical services for the time, including a twenty-six-bed hospital, an operating theater, radiology facilities, pharmacy, and a laboratory. The medical team consisted of a surgeon officer, a medical officer, and sixteen pharmacist's mates.

The dentist's office was also provided with the latest equipment and was capable of performing oral surgery as well as endodontics and periodontics. A dental officer and four assistants were embarked.

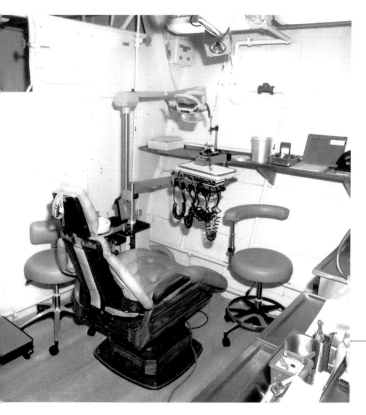

The dental surgery on board *Missouri* in August 1944.
NARA

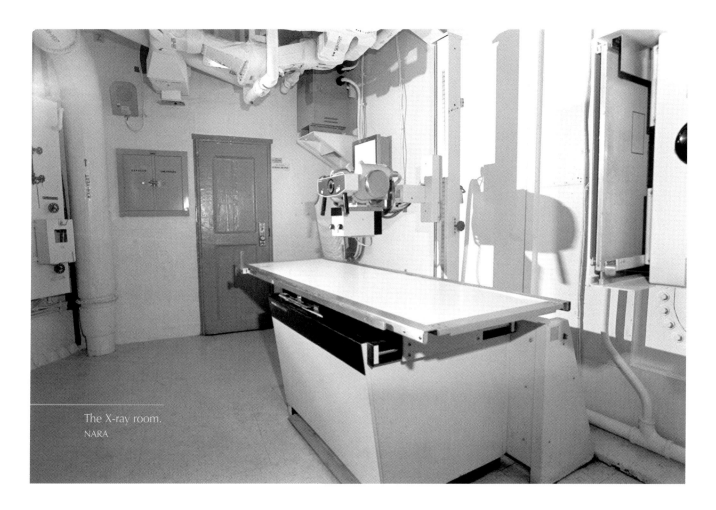

The X-ray room.
NARA

The pharmacy.
NARA

The operating theater.
NARA

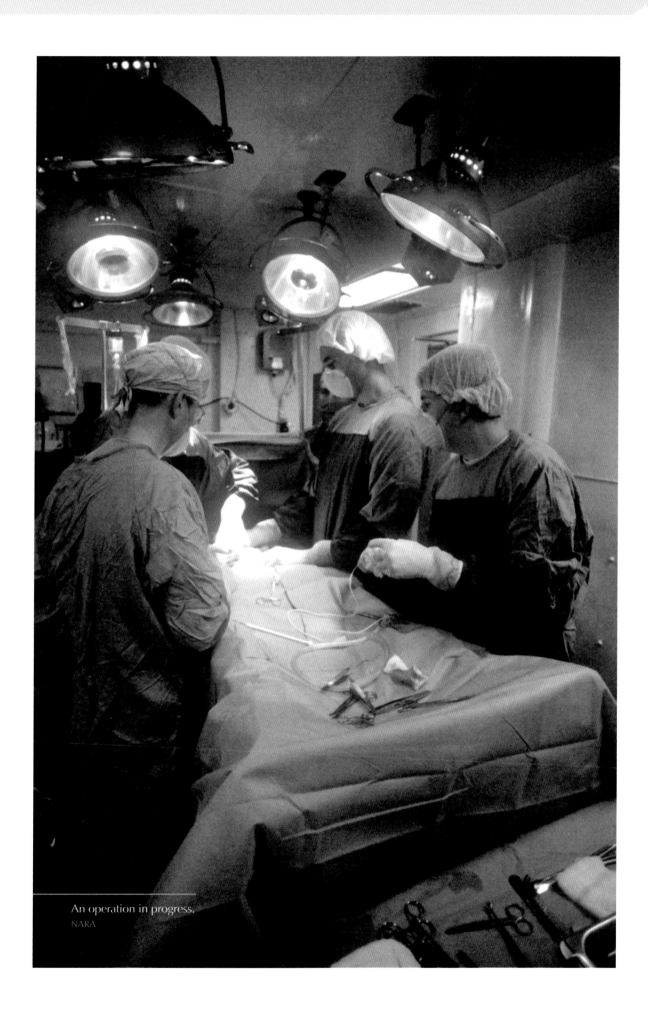

An operation in progress.
NARA

SIX

Paint and Camouflage

The *Iowa*-class units wore the following camouflage schemes throughout their careers:

- Ms 13—Haze Gray System, the standard livery for U.S. Navy warships. Vertical surfaces from boot-topping to top of superstructure masses, pole masts, yards, slender upper works above level of top superstructure masses painted a medium gray known as Haze Gray, 5-H; horizontal surfaces painted Deck Blue, 20-B; wood decking darkened to Deck Blue.
- Ms 21—Navy Blue System. Vertical surfaces painted Navy Blue 5-N; horizontal surfaces painted Deck Blue, 20-B; wood decking darkened to Deck Blue which was to be used in lieu of stain for that purpose.
- Ms 22—Graded System. Navy Blue 5-N applied to the hull to the height of the main deck edge at its lowest point. Haze Gray 5-H applied to all remaining vertical surfaces and all masts and small gear. Horizontal surfaces painted Deck Blue, 20-B. Wood decking darkened to Deck Blue, which was to be used in lieu of stain for that purpose.
- Ms 32-1B—Medium Pattern System. All exposed vertical surfaces painted a pattern of Light Gray 5-L and Black. All decks and horizontal surfaces painted a pattern of Deck Blue 20-B and Ocean Gray 5-0. "1B" stood for the first dazzle design for a battleship class.
- Ms 32/22D—Medium Pattern System. All exposed vertical surfaces painted a pattern of Light Gray 5-L, Ocean Gray 5-0, and Black. All decks and horizontal surfaces painted a pattern of Deck Blue 20-B and Ocean Gray 5-0. "22D" stood for the 22nd dazzle design from a destroyer-derived camouflage scheme.

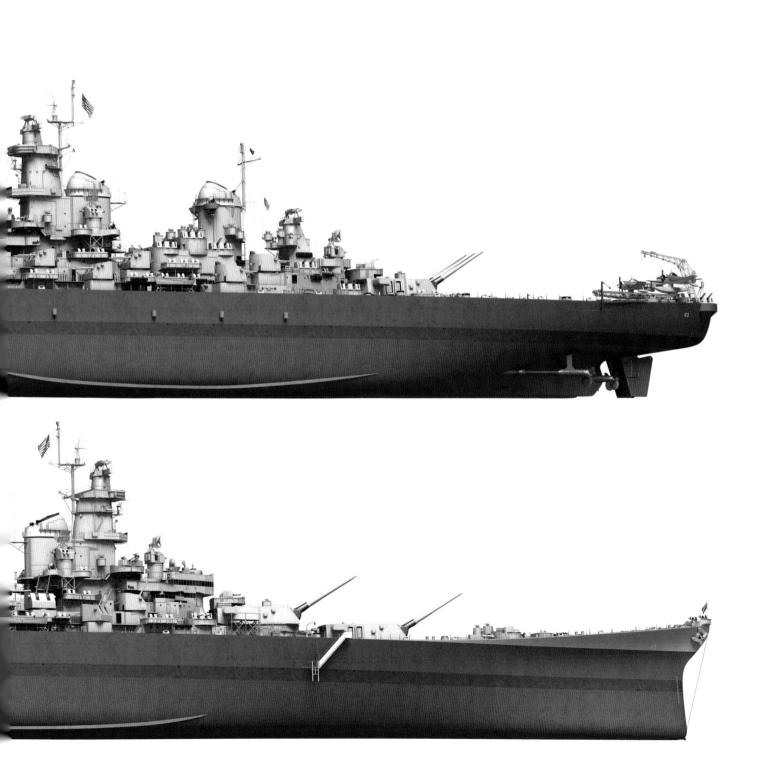

SEVEN

Commanding Officers

IOWA

Capt. John L. McCrea	2/22/1943
Capt. Allan R. McCann	8/_/1944
Capt. James L. Holloway	11/_/1944
Capt. Charles Wellborn	7/_/1945
Capt. Frederick Entwistle	11/_/1945
Capt. Raymond D. Tarbuck	7/_/1946
Capt. Thomas M. Stokes	4/_/1947
Capt. Edward W. Solomons	3/_/1948
Capt. William W. Jennings	8/_/1948
Capt. Bennett M. Dobson	1/_/1949
Capt. William Smedberg	8/25/1951
Capt. Joshua Cooper	7/_/1952
Capt. Wayne Loud	7/_/1953
Capt. William Bryson	9/_/1954
Capt. John W. Ailes	11/_/1955
Capt. Julian Becton	12/_/1956
Capt. Gerald Gneckow	4/28/1984
Capt. Larry Seaquist	7/_/1986
Capt. Fred Moosally	5/_/1988
Capt. John Moorse	5/23/1990

Admiral Bagget and Capt. Larry Seaquist, who commanded *Iowa* between July 1986 and May 1988.
USN

Capt. John L. McCrea, *Iowa*'s first captain, who commanded her between February 1943 and August 1944.
USN

Capt. Wayne Loud, who commanded *Iowa* between July 1953 and September 1954.
USN

NEW JERSEY

Capt. Carl Holden	5/23/1943
Capt. Edmund Wooldridge	1/26/1945
Capt. Edward Thompson	11/17/1945
Capt. Leon Huffman	8/5/1946
Capt. George Menocal	5/23/1947
Capt. Joseph Leverton	2/14/1948
Capt. David Tyree	11/21/1950
Capt. Francis McCorkle	11/17/1951
Capt. Charles Melson	10/20/1952
Capt. John Atkeson	10/24/1953
Capt. Edward O'Donnell	3/18/1955
Capt. Charles Brooks	5/31/1956
Capt. Edward Snyder	4/6/1968
Capt. Robert Peniston	8/27/1969
Capt. William Fogarty	12/28/1982
Capt. Richard D. Milligan	9/15/1983
Capt. Walter Glenn	9/7/1985
Capt. Douglas Katz	8/8/1987
Capt. Ronald Tucker	5/19/1989

MISSOURI

Capt. William Callaghan	6/11/1944
Capt. Stuart Murray	5/14/1945
Capt. Roscoe Hillenkoetter	11/6/1945
Capt. Tom Hill	5/31/1946
Capt. Robert Dennison	4/2/1947
Cdr. John B. Colwell	1/23/1948
Capt. James Thach	2/24/1948
Capt. Harold P. Smith	2/05/1949
Capt. William Brown	12/10/1949
Cdr. George E. Peckham	2/3/1950
Capt. Harold P. Smith	2/7/1950
Capt. Irving T. Duke	4/19/1950
Capt. George T. Wright	3/2/1951
Capt. John Sylvester	10/18/1951
Capt. Warner Edsall	9/4/1952
Cdr. James R. North	3/26/1953
Capt. Robert Brodie	4/4/1953
Capt. Robert Keith	4/1/1954
Cdr. James R. North	9/18/1954
Capt. Albert Lee Kaiss	5/10/1986
Capt. James Carney	6/20/1986
Capt. John Chernesky	7/6/1988
Capt. Albert Lee Kaiss	6/13/1990

Capt. Robert Peniston, who commanded New Jersey between August and December 1969.
USN

Capt. Edward Snyder, who commanded New Jersey between April 1968 and August 1969.
USN

Capt. Irving T. Duke, who commanded Missouri between April 1950 and March 1951.
USN

Capt. William Callaghan, who commanded Missouri between June 1944 and May 1945.
USN

Capt. Lee Kaiss, who commanded Missouri twice, first between May and June 1986, and then as her last CO from June 1990 to March 1992. Seen on the left is the mayor of San Francisco and later senator from California, Dianne Feinstein.
USN

WISCONSIN

Capt. Earl E. Stone	4/16/1944
Capt. John W. Roper	2/20/1945
Capt. Clark L. Green	12/18/1945
Capt. John M. Higgins	3/11/1947
Capt. Thomas Burrowes	3/3/1951
Capt. Henry C. Bruton	2/22/1952
Capt. Robert J. Foley	9/24/1952
Capt. Michael Flaherty	9/9/1953
Capt. Patrick Goldsborough	6/11/1954
Capt. Frederic Keeler	9/3/1955
Capt. John O. Miner	9/1/1956
Capt. Jerry M. Blesch	10/22/1988
Capt. David S. Bill	9/28/1990
Capt. Coenraad Schroeff	4/27/1991

Capt. Thomas Burrowes, who commanded *Wisconsin* between March 1951 and February 1952, seen here with the notorious senator from Wisconsin, Joseph R. McCarthy.
USN

Capt. Coenraad Schroeff, who served as *Wisconsin*'s last CO between April and September 1991.
USN

WISCONSIN: CONSTRUCTION PROGRESS

	MONTHS TO LAUNCH	TONNAGE ON SLIP (CUMULATIVE)	RIVETS DRIVEN (CUMULATIVE)	PERCENTAGE COMPLETION
January 1, 1941	35	0	0	0
October 1, 1941	26	862	11,529	4.3
January 9, 1942	23	1,700	64,000	7
April 1, 1942	20	2,452	140,000	10.8
July 1, 1942	17	3,900	243,000	18.4
January 1, 1943	11	9,400	534,000	35.8
April 1, 1943	8	14,200	633,000	43.8
October 1, 1943	2	23,300	1,025,000	67
May 31, 1944	—	28,236	1,211,183	100

Source: Muir, *The Iowa Class Battleships*, 147.

Iowa leads Task Unit 28.1.1 in May 1984. Note the two *Pegasus*-class hydrofoils on either side of her.
USN

Iowa carries out gunnery exercises on November 1, 1985.
USN

Iowa sailing with the frigate USS *Flatley* (FFG 21) in 1984.
USN

EIGHT
Battle Honors

Aside from Battle Stars, all the battleships were eligible for or received such campaign decorations as the World War II Victory Medal, the American Campaign Medal, the United Nations Korea Medal, the Korean Service Medal, National Defense Service Medal, Asiatic Pacific Campaign Medal, Vietnam Service Medal, Navy Occupation Service Medal, the Armed Forces Expeditionary Medal, the China Service Medal, Philippine Liberation Medal, Kuwait Liberation Medal, and the Navy Sea Service Deployment Ribbon.

Iowa
Nine Battle Stars for World War II
Two Battle Stars for the Korean War

Missouri
Three Battle Stars for World War II
Five Battle Stars for the Korean War
Three Battle Stars for the Gulf War

New Jersey
Nine Battle Stars for World War II
Four Battle Stars for the Korean War
Three Battle Stars for the Vietnam War
Three Campaign Stars for the Lebanon War and the Gulf War

Wisconsin
Five Battle Stars for World War II
One Battle Star for the Korean War
Two Battle Stars for the Gulf War

Iowa's award ribbons.
PHILIPPE CARESSE

New Jersey's award ribbons.
PHILIPPE CARESSE

BATTLE EFFICIENCY AWARD MARKINGS APPLIED NEAR THE CAMPAIGN DECORATIONS

E	Battle E: Award for the best ship handling, weapons employment, tactics, and ability to fulfill mission objectives (command and control). Only one award per squadron.	**CS**	Award for Combat Systems Excellence on board Aircraft Carriers
E	Excellence Award for the best Propulsion and Engineering Crews	**H**	Wellness Award for the best Health Promotion Activities. Units must apply for this award.
E	Excellence Award for the best Combat Information Centers (Surface Ships) / Award for Operations Excellence (Submarines)	**H**	Habitability Award
		M	Excellence Award for the best Medical Departments
E	Excellence Award for the best Supply Departments	**M**	Excellence Award for the best Medical Departments
E	Excellence Award for Maritime Warfare Excellence (Surface Ships) / Excellence Award for Aviation Maintenance Excellence (Aircraft Carriers)	**N**	Award for Navigation Excellence
		N	Award for Navigation Excellence
		T	Excellence Award for Tactical Proficiency
		R	Award for Repair Excellence
E	Excellence Award for the best Air Departments on board Aircraft Carriers	**D**	Excellence Award for the best Deck Departments
F	Award for Fire Control Excellence	**W**	Excellence Award for the best Weapons Departments on board Aircraft Carriers
C	Excellence Award for the best Communication Departments	**D**	Dental Award
		DS	Excellence Award for Deep Submergence
A	Award for Antisubmarine Warfare Excellence	**S**	Excellence Award for Strategic Operations
DC	Excellence Award for the best Damage Control Crews	⚓	Navigation Award
		⚓	Deck Seamanship Award

Source: David Way

NINE

Refitting and Conversion Projects

From the time they entered service the ships of the *Iowa* class were relegated to the role of antiaircraft batteries intended for the protection of aircraft carriers, although their heavy guns played a significant role in bombarding shore targets to cover troop landings. World War II established the aircraft carrier as the centerpiece of every battle fleet, and it is no surprise that from 1942 onward all warships under construction from light cruiser and up were susceptible to conversion as carriers. Accordingly, in 1944 the Bureau of Ships (BuShips) began taking a particular interest in *Kentucky* (BB 66), then nearing completion and given that her hull design lent itself to conversion to a carrier similar to the *Essex* class. This project, however, did not survive the end of the war and it was the light cruisers of the *Cleveland* class that provided the basis for the *Independence*-class escort carriers (CVL 22 to CVL 30).

In the early 1950s it was proposed to convert *Kentucky* to a guided-missile battleship (BBG 1). With a budget of $15–30 million, the vessel would have retained her two forward 16-inch turrets together with her 5-inch guns but would have been equipped with two twin Terrier missile launchers on her fantail. There was to have been stowage for two hundred missiles.

In 1955 proposals were put forward for a guided-missile monitor battleship (BB MG) under which the *Iowa*-class ships would have lost their guns in favor of four twin Talos missile launchers (400 missiles embarked), twelve Tartar missile launchers (2,400 missiles embarked), and a vertical missile launcher for the Jupiter SM-78S weapon.

Also in 1955 two draft projects envisioned the *Iowa*-class ships as fleet replenishment ships. The entire armament would be landed with the exception of four 5-inch mounts and the superstructure removed together with the two outer propeller shafts, thereby restricting speed to 26 knots. These plans were quickly abandoned, however, in view of the limited oil capacity.

Further projects followed as part of the 1957 program, with scheme I proposing a battleship with two forward turrets, four 5-inch mounts, and twin Talos Mk-7 missile launchers. Scheme II followed the same arrangement but with an additional Talos launcher.

In the late 1950s a guided-missile battleship (BBG) program was proposed known as SCB 173, with a budget of $85 million. Scheme I of this program provided for retention of the two forward turrets and the fitting of two twin Talos launchers, four twin Tartar launchers, a launcher for the Regulus II, twelve Polaris RGM27 silos, and an ASROC launcher. Scheme II was essentially the same project but with all 16-inch turrets landed.

In 1962 it was proposed to look into the conversion of the *Iowa* class to command and landing support ships. The two forward turrets were to be retained but the superstructure cleared on either side to accommodate fourteen LCM6 landing craft. The vessel would be capable of embarking thirty helicopters together with 1,800 troops and their equipment.

The 1964 program was certainly the most original since it envisioned the transformation of a unit of the class into a satellite launcher. The entire superstructure and armament would be removed and replaced with dedicated facilities for a new advanced gun satellite launch (AGSL) system. The firing area was located on the fantail with the system capable of launching two Atlas D CGM-16 rockets. Two propeller shafts would be removed, restricting speed to 27 knots, but there is no information on protection or full-load displacement. It is unlikely, though, that any such vessel could have entered service since the U.S. Air Force of the day was determined to maintain control over the use of missiles and satellites and would never have agreed to a Navy vessel serving as a launch platform.

In all cases with the exception of the fleet replenishment ships and the satellite launcher, the displacement and speed were unchanged at 52,000 tons full load and 32.5 knots, respectively, with a large proportion of the armor redistributed or removed.

We now know that the various conversions planned for the *Iowa* class were much too expensive for vessels that already had respectable careers behind them and were not truly adapted to their intended tasks. Moreover, the electronic components planned for installation were much too sensitive to withstand the firing of the 16-inch guns.

TEN

Previous Ships of the Name

The first vessel in the U.S. Navy to bear the name *Iowa* was a steam frigate launched at Boston Navy Yard in 1864 as the *Ammonoosuc*. She displaced 3,850 tons, was 335 feet long and 44 feet 4 inches in the beam and carried ten 9-inch smoothbore, three 60-pounder rifled, and two 24-pounder smoothbore guns. Fitted with Morgan Iron Works steam engines, a series of steam trials carried out in the summer of 1868 found her capable of 17.11 knots, the highest sustained speed ever attained by a ship to that time, but *Ammonoosuc* was never commissioned, due probably to the fact that so much of her hull was taken up by machinery. Laid up in the same yard, she was renamed *Iowa* on May 15, 1869, being eventually sold on September 27, 1883.

USS *IOWA* (BB 4)

The sole unit in her class, the battleship *Iowa* was laid down by William Cramp & Sons Shipbuilding Company of Philadelphia on August 5, 1893, launched on March 28, 1896, and commissioned on June 16, 1897. She displaced 11,528 tons and was 362 feet 5 inches long and 72 feet 3 inches in the beam. Two triple-expansion engines produced 11,000 ihp for a speed of 17 knots. Armament was composed of four 12-inch, eight 8-inch, six 4-inch, and twenty 6-pounder guns. Protection consisted of a 14-inch armor belt, with 17 inches on the turrets and 10 inches on the conning tower.

Iowa participated in the Spanish-American War, bombarding San Juan, Puerto Rico, on May 12, 1898, and blockading Santiago de Cuba sixteen days later. On July 3, she was present at the Battle of Santiago, during which she suffered two hits. Nonetheless, she had a hand in the destruction of the destroyer *Plutón* before taking up the pursuit of the cruiser *Vizcaya*, which was forced to run herself aground. After the cessation of hostilities she rejoined the Pacific Fleet before being assigned to the South Atlantic Squadron. She was relegated to the reserve on June 6, 1907, though not before the future admiral Raymond A. Spruance joined her the previous year.

Now fitted with a cage mast, *Iowa* was recommissioned on May 27, 1910, and on May 13, 1911, assisted in rescuing passengers from the sunken liner *Merida*. She was disarmed on May 27, 1914, however, and spent World War I as a training ship, being renamed *Coast Battleship No. 4* on April 30, 1919. The former *Iowa* became the U.S. Navy's first remote-controlled target ship and was sunk in the Gulf of Panama on March 23, 1923.

Mention should also be made of the battleship bearing the name with hull number BB 53 that was cancelled together with the entire *South Dakota* class under the terms of the Washington Naval Treaty in February 1922. She was sold for scrapping on November 8, 1923, when 31.8 percent complete.

The battleship *Iowa* (BB 4) drydocked at Brooklyn.
LIBRARY OF CONGRESS

The battleship *New Jersey* (BB 16) in 1918 with her unique camouflage.
USN

The battleship *Missouri* (BB 11) at full speed c. 1905. USN

USS *NEW JERSEY* (BB 16)

The battleship *New Jersey* was a unit of the *Virginia* class. She was laid down at the Fore River Shipyard at Quincy, Massachusetts, on April 2, 1902, launched on November 4, 1904, and commissioned on May 12, 1906. She displaced 13,561 tons and was 441 feet 3 inches long and 76 feet 3 inches in the beam. Two triple-expansion engines produced 19,000 ihp for a speed of 19 knots. The armament was composed of four 12-inch, eight 8-inch, and twelve 6-inch guns. Protection consisted of an 11-inch armor belt, 12-inch turrets, and a 9-inch conning tower.

In September 1906 *New Jersey* received a visit from President Theodore Roosevelt at Oyster Bay, New York, after which she deployed to Cuba between September 21 and October 13 to protect U.S. interests during the second occupation of Cuba.

On December 16, 1907, she joined the so-called Great White Fleet on its world cruise, which lasted until February 22, 1909, before being decommissioned in Boston between May 2, 1910, and July 15, 1911. She served as a training ship in the summers of 1912 and 1913 until ordered to the Mexican coast, where a landing by U.S. Marines captured the port of Veracruz in the aftermath of the Tampico Affair. *New Jersey* was then ordered to Santo Domingo and Haiti. During World War I she served as a gunnery training ship and receiving vessel for draftees in Chesapeake Bay, repatriating more than five thousand U.S. troops from France after the Armistice. She was disarmed on August 6, 1921, and sunk off Diamond Shoals on September 5, 1923, by aircraft commanded by Gen. William "Billy" Mitchell.

USS *MISSOURI* (BB 11)

The first *Missouri* was a 229-foot, 3,220-ton side-wheel frigate laid down at the New York Navy Yard in 1840 and launched in January the following year. Among the earliest steam-powered warships in U.S. Navy service, in 1842 she became the first American naval vessel to cross the Atlantic under steam. It was the start of a planned voyage to the Far East to secure a trade agreement with China. Unfortunately, her career came to a sudden end at Gibraltar on August 26, 1843, after the accidental rupture of a container of turpentine in a storeroom set her ablaze and led to the detonation of a powder magazine.

The next *Missouri* was a side-wheel ironclad steamer laid down for the Confederate Navy at Shreveport, Louisiana, and launched in April 1863. Intended for use as a ram against Union ships, she instead served as a troop transport and engaged in mining activities. Her crew turned her over to the Union in June 1865, two months after the Confederate surrender. Because the ship was in poor material condition, the U.S. Navy never commissioned her for service and instead auctioned her for scrapping.

The battleship *Missouri* was a unit of the *Maine* class. She was laid down at Newport News Shipbuilding & Drydock Company in the city of that name in Virginia on February 7, 1900, launched on December 28, 1901, and commissioned on January 12, 1903. She displaced 12,200 tons and was 393 feet 11 inches long and 72 feet 3 inches in the beam.

Two triple-expansion engines produced 16,000 ihp for a speed of 18.15 knots. Her armament was composed of four 12-inch, sixteen 6-inch, six 3-inch, and eight 3-pounder guns. Protection consisted of an 11-inch armor belt, 12-inch turrets, and a 10-inch conning tower.

Initially assigned to the North Atlantic Fleet, on April 13, 1904, *Missouri* suffered a flareback incident during a gunnery exercise that ignited three propellant charges in the after turret, left thirty-six men dead, and came close to destroying the ship, but a greater disaster was averted by orders to flood the magazines. On the bridge during this incident was midshipman and future admiral William F. Halsey. On January 1907, *Missouri* rendered assistance to victims of the earthquake that struck Kingston, Jamaica, and on December 16 that year joined the Great White Fleet on its famous world cruise, which ended on February 22, 1909.

Missouri was placed in reserve in Boston on May 1, 1910, but recommissioned on June 1, 1911, deploying to Cuba with a contingent of Marines in June 1912. Decommissioned at Philadelphia three months later, she was nonetheless reactivated between 1914 and 1917, during which she served as a training ship for midshipmen. In 1919 she began a series of voyages to repatriate the first of 3,278 U.S. troops from Brest. *Missouri* was stricken on September 8, 1919, and sold for scrapping on January 26, 1922.

USS *WISCONSIN* (BB 9)

The battleship *Wisconsin* was a unit of the *Illinois* class. She was laid down at the Union Iron Works in San Francisco on February 9, 1897, launched on November 26, 1898, and commissioned on February 4, 1901. She displaced 12,350 tons and was 373 feet 10 inches long and 72 feet 3 inches in the beam. Two triple-expansion engines produced 10,000 ihp for a speed of 16 knots. Her armament was composed of four 13-inch, fourteen 6-inch, sixteen 6-pounder, and six 1-pounder guns. Protection consisted of a 16.5-inch armor belt, 14-inch turrets, and a 10-inch conning tower.

Having completed her shakedown cruise in Magdalena Bay off Baja California in March 1901, *Wisconsin* embarked on a cruise of the U.S. West Coast with numerous ports of call. In October she sailed for Hawaii and reached Samoa on the 26th, Tutuila in American Samoa on November 5, and Apia at the end of that month before returning to Pearl Harbor. This was followed by visits to Acapulco, Callao, Valparaíso, and Acapulco once again on February 26, 1902. In March 1903 she carried out exercises in Magdalena Bay prior to heading north with stops at Coronado, San Francisco, and Port Angeles before making her way to Puget Sound Naval Shipyard for refitting between June 4 and August 11. In September she became the flagship of Rear Adm. Sillas Casey III commanding the Pacific Squadron. Casey acted as a mediator during Colombia's Thousand Days' War while *Wisconsin* lay at Panama (September 30–November 22). Returning to San Francisco on December 6, 1903, she went into refit until May 19, 1903, after which she was assigned to the Asiatic Fleet, visiting Kōbe, Yokohama, Nagasaki, Yokosuka, Amoy, Shanghai, Chefoo, Nanking, and Taku while serving first under Rear Adm. Yates Stirling and then Rear Adm. Philip H. Cooper.

Wisconsin was decommissioned at Bremerton on November 16, 1906, but returned to service on April 1, 1908. Rejoining the Atlantic Fleet, she made stops at Rockport, Provincetown, New York City, Portsmouth, Newport, and so forth. In the spring of 1910 she passed into the reserve before joining the Naval Academy training squadron in 1915. On April 23, 1917, she was assigned to the Coast Battleship Patrol Squadron and carried out her final cruise in Cuban waters in the summer of 1919 before being decommissioned on May 15, 1920. She was sold for scrapping on January 26, 1922.

The battleship *Wisconsin* (BB 9) in 1903.
USN

Kentucky under construction at Norfolk Naval Shipyard on February 4, 1946.
USN

The hull of the *Kentucky* on March 26, 1946.
USN

ELEVEN

The Sad End of the *Illinois* and *Kentucky*

Two units of the *Iowa* class were never completed although construction was authorized by Congress on July 19, 1940. The orders for *Illinois* and *Kentucky* were placed with Philadelphia Naval Shipyard and Norfolk Naval Shipyard, respectively, but the commencement of the work was much delayed.

Illinois was laid down on January 15, 1945, but the work was suspended three days after the Japanese surrender with the ship only 22 percent complete. It was briefly contemplated to expend her as a target in nuclear tests, but the costs of completing her to the point she could be towed were too high and it was decided to scrap her on the slip from September 1958.

Kentucky was laid down on December 6, 1944, with construction halted in August 1946. Numerous conversion programs were subsequently considered, particularly that to an antiaircraft battleship, a project known as SCB 19. *Kentucky* was launched 73 percent complete on January 20, 1950. Her bow, weighing 120 tons, was transferred to *Wisconsin* after her collision with the destroyer *Eaton* on May 6, 1956.

In 1955 earlier plans were revived to convert *Kentucky* to a guided-missile battleship (BBG) with only turrets 1 and 2 retained and Regulus II missiles embarked. The antiaircraft armament was to be composed of Terrier, Tartar, and Talos missile launchers together with six single 5-inch mounts and ten twin 3-inch turrets. It was subsequently intended to embark sixteen Polaris missiles. This project was included in the 1958 budget ($200 million) but did not survive. *Kentucky* was stricken on June 9, 1958, and sold for scrapping to Boston Metals Company of Baltimore for the sum of $1,176,666 on October 31. Her boilers and turbines were utilized in the fast combat support ships *Sacramento* (AOE 1) and *Camden* (AOE 2).

TWELVE

Preservation and Legal Status

Between 1998 and 2012 all the *Iowa*-class battleships were transferred to the ownership of a private non-profit entity or, in the case of *Wisconsin*, by the host city where she lies (Norfolk, Virginia). These contracts are on public record and although there are minor differences between them the substance is essentially the same, particularly with respect to the preservation of each vessel and its potential return to government service.

In the case of *Iowa*, per the contract signed with the U.S. Navy on April 30, 2012, the ship is owned by the Pacific Battleship Center (PBC), a 501(c)(3) nonprofit organization, and is subject to the conditions set forth therein. These include the following provisions:

- PBC accepts title to the vessel and assumes all financial responsibility.
- There will be no future cost to the federal government.
- PBC must operate the vessel as a stationary ship museum/memorial.
- PBC must maintain its nonprofit status.

Furthermore, the contract states that the PBC must maintain the vessel and its appurtenances in a condition satisfactory to the Secretary of the Navy. The latter is not particularly well defined in the case of *Iowa*, but it does include maintaining the ship in towable condition, providing ongoing corrosion and dehumidification protection, and providing the government with reasonable access to the ship for annual inspection purposes. With respect to *Wisconsin*, the following provisions were imposed:

- No alterations whatsoever are permitted to anything affecting the military utility of the ship.
- The battleship is to be maintained in her current state through continuous use of cathodic protection, dehumidification systems, and other preservation methods.
- Sufficient guns and shells are to be preserved to permit possible reactivation of the vessel.
- The U.S. Navy is to prepare plans for rapid reactivation of *Wisconsin* in the event of a national emergency.

There are additional provisions within the PBC contract that reference historic preservation, potential disposal of the vessel, and financial assignment, all of which exist in the other contracts in various forms. In particular, paragraph 10 of the PBC contract addresses the termination, transfer, or disposal of the vessel, stating as follows: "The Secretary of the Navy may terminate the contract and retake possession of the Vessel as-is, where-is, should the Secretary determine that the Vessel is needed for reactivation in the event of a national emergency." The simple answer to the oft-raised question as to the possibility of reversion of *Iowa* to federal control is that if the Navy wants her they can take her away.

Kentucky on December 21, 1949. Beyond her is the carrier *Coral Sea* (CVA 43).
USN

Kentucky seen minus her bow at Norfolk in 1956.
USN

Kentucky makes her last voyage as she is led away for demolition at Baltimore on February 6, 1959.
USN

Illinois under construction at Philadelphia Naval Shipyard.
USN

New Jersey carries out a Tomahawk firing off the coast of California on May 10, 1983.
USN

New Jersey opens up with her main battery against Kaesong, Korea, on January 1, 1953. Note the parabolic array for the SPS-6 radar above the main rangefinder.
USN

PART II
Service Careers

THIRTEEN

The Career of the Battleship USS *Iowa* (BB 61)

Iowa Tested in the Pacific War

The keel of *Iowa* (BB 61) was laid at the New York Naval Shipyard on June 27, 1940. The first rivet was placed by Rear Adm. Clark H. Woodward, head of the Bureau of Naval Construction. The ship was launched on August 27, 1942, with a hull weight of 36,346 tons. Her sponsor was Mrs. Ilo Wallace, wife of U.S. vice president Henry Wallace.

Work progressed smoothly and the superstructure soon displayed its elegant silhouette. The main turrets each required more than a month's work to complete, including installation of the faceplate, which alone weighed one hundred tons. Despite the priority given to aircraft carrier construction, *Iowa* was completed with commendable speed in thirty-two months. She was commissioned on February 22, 1943, under Captain McCrea, and by tradition no time was lost in bringing on board the silver service carried by her earlier namesake and veteran of the Spanish-American War of 1898. Within a few months of joining the U.S. Navy, the crew had nicknamed their ship "The Mighty I."

The battleship began her shakedown cruise two days later in Chesapeake Bay. The period before the ammunitioning of the ship required the finishing touches to her construction, adjustment of the compasses, and completion of speed, fuel consumption, and steam trials, as well as fire, flooding, and man overboard drills together with trials of the Kingfisher catapults.

Iowa's triple bottom in the course of assembly on September 30, 1940.
NARA

A view looking forward showing progress on the construction of the hull on December 30, 1940.
NARA

The barbette of turret 3 and eight boilers in place on June 27, 1941.
NARA

Iowa under construction at the New York Naval Shipyard on October 3, 1941.
NARA

The superstructure of *Iowa* taking shape on July 3, 1942.
NARA

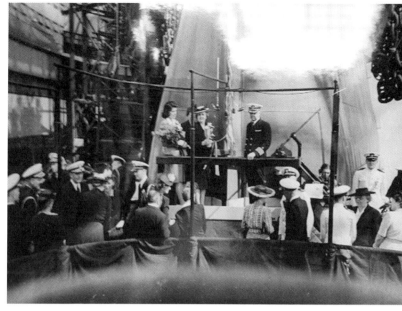

Mrs. Ilo Wallace, wife of U.S. vice president Henry Wallace, about to launch the most powerful battleship in the U.S. Navy on August 27, 1942.
NARA

Iowa's massive bow shortly before launching.
NARA

Iowa slides down the ways on August 27, 1942.
NARA

Iowa seen shortly after launching and waiting to be towed to the fitting-out basin.
USN

Iowa being maneuvered into position by tugs.
NARA

A crowd has gathered to watch *Iowa* being brought into dry dock.
NARA

Turret 2 under construction.
USN

The fantail with work proceeding on turret 3.
USN

The forecastle with work on the turrets nearing completion.
USN

Iowa tied up at the fitting-out basin.
USN

Iowa's commissioning ceremony on February 22, 1943.
USN

Iowa at Bayonne Naval Shipyard, New Jersey.
NARA

The 20-mm mounts in place in the eyes of the ship.
NARA

The 40-mm and 20-mm mounts on the forecastle.
NARA

Iowa's original three-level conning tower seen from the roof of turret 2. *Iowa* was fitted as a Task Force flagship while her three sisters were completed as Division flagships. Consequently, she was unique in that the 17.3-inch armor protection of the conning tower not only covered the 05 Level Fire Control Station and 04 Level Bridge, but continued down to the 03 Level Flag Bridge. To avoid obstructing the admiral's line of sight from the 03 Level viewing slits, *Iowa* mounted three 20-mm guns rather than the single quad 40-mm mount fitted in her three sisters.
USN

Iowa lying at anchor on April 4, 1943, while running trials.
NARA

Iowa seen on March 28, 1943, shortly before leaving the dry dock at Bayonne Naval Shipyard.
USN

On July 16, *Iowa* struck a submerged rock or wreck while entering Casco Bay, Maine, at 12 knots and grounded in a shallow, meandering channel. Inspection by divers revealed the damage to be serious, with rivets missing on the port side and several breaches in the hull along 252 feet of her length. The incident damaged sixteen fuel bunkers, and eighteen steel plates on the hull had to be replaced. *Iowa* was ordered to a floating dock in Boston, where she was immobilized for four weeks. Admiral Bryant, commanding the Atlantic Fleet Battleship Squadron, demanded that McCrea be relieved for this navigational error, but after an inquiry President Roosevelt personally intervened in favor of his former aide-de-camp, who had been at his side during the first year of the war.

Operational once more from August 27, *Iowa* headed for Argentia in Newfoundland in readiness to intercept the German battleship *Tirpitz* should she sortie into the Atlantic. With a full-load displacement of 53,500 tons, eight 15-inch guns, and a maximum speed of 30.8 knots, *Tirpitz* would have been a serious proposition for the newly commissioned *Iowa*. But she was temporarily put out of action by British midget submarines in a Norwegian fjord on September 22, a situation that allowed *Iowa* to return to Norfolk on October 25 for refitting and to prepare for a special assignment.

An aerial view of *Iowa*'s midships section showing the after stack, antiaircraft armament, and 5-inch mounts.
NARA

Iowa being maneuvered in the port of New York.
USN

Iowa at Boston on August 27, 1943.
USN

The German battleship *Tirpitz*, whose threatened appearance in the Atlantic kept *Iowa* anchored at Argentia, Newfoundland, for two months in the fall of 1943.
USN

Iowa after her stay at Argentia and being readied for one of her most important missions: conveying President Roosevelt across the Atlantic to attend the Tehran Conference in November 1943.
NARA

President Roosevelt delivers a speech in *Iowa* on coming on board the ship in Chesapeake Bay on November 12, 1943.
USN

November 11, 1943, found *Iowa* anchored in Chesapeake Bay. At 0916 the following day she welcomed on board President Franklin Delano Roosevelt for passage across the Atlantic to attend the Tehran Conference. With him were members of his staff including the chief of staff to the commander in chief, Adm. William D. Leahy; the chief of staff of the Army, George C. Marshall; the Chief of Naval Operations, Admiral Ernest King; the commanding general of the U.S. Army Air Forces, Henry H. "Hap" Arnold; presidential emissary and adviser Harry Hopkins; and many other senior members of the U.S. high command. The destroyers *William D. Porter* (DD 579), *Young* (DD 580), and *Cogswell* (DD 651) constituted the escort of what became Task Group 27.5.

After various maneuvers in Hampton Roads, *Iowa* returned to anchor and was joined by the oilers *Escalante* (AO 70) and *Housatonic* (AO 35) to fill her fuel tanks. Meanwhile, Captain McCrea had brought on board his dog Victory, known as Vicky, who quickly took up residence in the president's quarters and even slept on his bed. She remained in the ship with her master until August 1944 having covered 205,000 miles in the rank of M1C (mascot 1st class).

Capt. John L. McCrea and his dog, Vicky, who served on board with him until August 1944.
PRIVATE COLLECTION

At 1206 on the 13th, TG 27.5 weighed anchor and steered a course of 105 degrees while zigzagging at 25 knots in heavy seas, the entire crew at general quarters and the president's flag unhoisted for security reasons. At 0800 the following day an antiaircraft exercise was successfully carried out against balloon targets east of Bermuda. But the voyage was by no means uneventful. Providing close escort was the destroyer *Porter*, which had been in a collision with one of her sisters on sailing from Norfolk two days earlier, damaging her with her anchor. Having closed with *Iowa*, one of her depth charges was inadvertently dropped and exploded, causing the entire formation to maneuver at speed believing it had come under submarine attack. Shortly after, a failure in one of *Porter*'s firerooms obliged her to reduce speed until another boiler was brought online.

Iowa under way in the Atlantic with President Roosevelt embarked. Note the elevators fitted immediately forward of No. 53 and (just visible) No. 55 5-inch mounts to transfer the wheelchair-bound president from deck to deck.
NARA

An antiaircraft firing and torpedo launch exercise were subsequently planned. The former was carried out successfully against a number of target balloons released by *Porter*, but the torpedo exercise came close to disaster. Torpedomen Lawton Dawson and Tony Fazio of the *Porter* were responsible for removing the primers from the torpedoes, but Dawson failed to remove it from tube 3. Consequently, when Lt. Cdr. Wilfred A. Walter passed the order for the exercise to commence, the characteristic sound of an armed torpedo being launched was heard, in this case heading straight toward *Iowa*. A lieutenant immediately asked Walter if he had given the order to fire this torpedo. When the latter answered in the negative it was decided to warn the battleship by signal lamp, but the effort came to nothing and the decision was made to break radio silence. *Iowa*'s radio operator initially bridled at the failure to comply with security instructions and requested confirmation of identity. The operator's urgent tone finally alerted Captain McCrea to the danger and *Iowa* took evasive action at over 29 knots before the torpedo detonated four minutes after launching. Meanwhile, President Roosevelt had asked his security detail to move his wheelchair onto the fantail to view the spectacle. The danger passed, questions were asked in *Iowa* as to whether Roosevelt and his chiefs of staff had been the target of some assassination plot, and the hapless *Porter* was detached to Bermuda for a court of inquiry to be held. During the latter, Chief Torpedoman Lawton Dawson acknowledged his error and was sentenced to fourteen years of hard labor, but Roosevelt interceded to demand that no one be punished for this incident. Walter remained in command of the *Porter* and eventually became a rear admiral, but the dossier on the incident remained classified until the end of the war.

The hapless destroyer *William D. Porter* (DD 579), which inadvertently launched a torpedo against *Iowa* during her presidential crossing of the Atlantic.
NARA

At 0915 on the 15th the escort was replaced by the destroyers *Hall* (DD 583), *Macomb* (DD 458), and *Halligan* (DD 584). At 1451 the following day the escort carrier *Block Island* (CVE 21) and three destroyers joined to provide air cover. On the 17th it was the turn of the destroyers *Ellyson* (DD 454), *Rodman* (DD 456), and *Emmons* (DD 457) to escort *Iowa*, the squadron being reinforced on the 19th by the arrival of the light cruiser *Brooklyn* (CL 40) and five destroyers, of which three were British.

The Strait of Gibraltar was traversed between 2111 and 2134, during which time the ships were caught in the beam of several searchlights, obliging the destroyers to lay a smokescreen. At 0715 on the 20th *Iowa* sighted Oran and anchored at Mers el Kébir at 0809. Roosevelt disembarked a half hour later preparatory to onward travel by air to participate in the historic conference.

Meanwhile, the commander in chief of the British Mediterranean Fleet, Admiral Sir Andrew Cunningham, had cause to fear for the safety of the battleship. Two months earlier the Luftwaffe had sunk the Italian battleship *Roma* and damaged *Italia* (ex-*Littorio*) in glider bomb attacks in the Mediterranean. A similar attack was a distinct possibility if the German intelligence service (Abwehr) learned of the presence of the flagship of the U.S. Navy at Mers el Kébir. *Iowa* therefore sailed that same day for the Brazilian port of Bahia, crossing the equator for the first time in the process. Reaching Bahia on the 29th, she was under way the following day to refuel at Freetown, Sierra Leone, from which she departed on December 6 for Dakar, Senegal, as a unit of Task Group 27.5, there to await the arrival of the president and his staff for the return voyage to America.

After embarking an additional 2,600 tons of fuel oil, *Iowa* was ready to sail at short notice, and no sooner had Roosevelt reached Dakar on the 9th than she weighed anchor at 0900 and sailed under escort by *Ellyson*, *Rodman*, and *Emmons*. At 2313 on the 15th, *Iowa*'s radar allowed her to avoid a certain collision with a darkened steamer in the Atlantic, and the following day she picked up a pilot to enter Chesapeake Bay in safety. *Iowa* anchored at the mouth of the Potomac, and Roosevelt disembarked at 1759 on the 16th after congratulating the crew. Her four crossings of the Atlantic had taken her 16,161 miles at an average speed of 23.6 knots without the slightest mechanical mishap. That same month *Iowa* attained 31.9 knots on a displacement of 56,928 tons while developing 221,000 shp.

Iowa arriving at Mers el Kébir on November 20, 1943. The elevator towers specially installed for the president's use are circled.
NARA

Iowa operating off the Marshall Islands on January 24, 1944. The battleship *Indiana* is steaming off her starboard quarter.
USN

Iowa and *Indiana* under way in the Pacific in August 1944. USN

On January 2, 1944, *Iowa* became flagship of the 7th Battleship Division of the Fifth Fleet in the Pacific (Task Group 58.3), serving under Rear Adm. Olaf M. Hustvedt. *New Jersey* was also part of this Task Group, being assigned to Adm. Marc A. Mitscher's Task Force 58. After sailing from Hampton Roads, *Iowa* traversed the Panama Canal with *New Jersey* on the 7th and the two ships briefly stopped at Balboa before sailing for the Gilbert and Ellice Islands the following day. After crossing the equator on the 13th they reached Funafuti Atoll on the 22nd, where they joined Vice Adm. Raymond Spruance's Fifth Fleet. Awaiting them were the battleships *North Carolina* (BB 55) and *Washington* (BB 56) together with a variety of aircraft carriers and escort vessels.

Iowa and *New Jersey* aside, Task Group 58.3 consisted of the carrier *Bunker Hill* (CV 17), the escort carriers *Cowpens* (CVL 25) and *Monterey* (CVL 26), and the heavy cruiser *Wichita* (CA 45), together with the destroyers *Wilson* (DD 408), *Izard* (DD 589), *Conner* (DD 582), *Bell* (DD 587), *Charrette* (DD 581), *Burns* (DD 588), *Bradford* (DD 545), *Brown* (DD 546), and *Cowell* (DD 547). Shortly after their arrival Rear Admiral Hustvedt and the commanders of *Iowa* and *New Jersey* came on board *Bunker Hill* to report to the Task Group commander Rear Adm. Forrest Sherman, who lost no time in telling them he had no interest in knowing whether they could fire their 16-inch guns, only in the effectiveness of their antiaircraft defense. The danger came from the sky, not from enemy battleships.

After a night of replenishment and repairs to the SG surface-search radar, the ship sailed for Kwajalein Atoll in connection with Operations Flintlock and Catchpole against the Marshall Islands, including the conquest of Kwajalein and Majuro Atolls and Eniwetok. *Iowa* and TG 58.3 waited at readiness off Bikini during the attacks but no Japanese aircraft appeared. The Marshall Islands campaign was a success and the Task Group sailed for Majuro on February 4. Five days later Vice Adm. Raymond Spruance hoisted his flag in *New Jersey* and established Task Group 50.9, composed in addition of *Iowa*, the heavy cruisers *Minneapolis* (CA 36) and *New Orleans* (CA 32) and the destroyers *Burns* (DD 588), *Izard* (DD 589), *Charrette* (DD 581), and *Bradford* (DD 545).

Iowa entering Majuro in the Marshall Islands on February 4, 1944.
USN

Crewmen scrubbing the forecastle under the two forward 16-inch guns in 1951.
USN

Iowa anchored in an atoll in the Pacific.
USN

The Japanese destroyer *Maikaze*.
USN

The Japanese destroyer *Nowaki*.
USN

On the 16th, the Task Group sailed toward Truk to bombard the Japanese naval base (Operation Hailstone). Weather conditions were excellent with fifteen miles of visibility, sea state 2, and a fresh breeze. The first air attack came around midday, when a bomb dropped by a Zero exploded thirty yards off *Iowa*'s starboard bow. The following day *New Jersey*'s reconnaissance aircraft reported enemy vessels twenty-five miles to the northwest. These turned out to be the light cruiser *Katori*, the destroyers *Nowaki* and *Maikaze*, and the armed trawler *Shonan Maru No. 15*. The two battleships went on to 32.5 knots and sighted the enemy at a range of 33,000 yards. The Japanese squadron had already been attacked by F6F Hellcats and TBF Avengers from the carriers *Yorktown* (CV 10), *Intrepid* (CV 11), *Essex* (CV 19), *Bunker Hill* (CV 17), and *Cowpens* (CVL 25). *Katori* was damaged and her speed reduced to a crawl by the time "The Mighty I" opened fire on her at 1527. The fifth salvo scored a heavy hit on *Katori*'s starboard side but she still managed to unleash three torpedoes at *Iowa*, which were easily evaded. Seven hits, however, had ravaged the starboard side of the cruiser, opening holes five feet in diameter, while she had also been struck below the waterline on the port side. After thirteen minutes *Katori* sank by the stern in position 07°45′ N, 151°20′ E. Assisted by *Minneapolis* and *New Orleans*, *New Jersey* destroyed *Maikaze* and the trawler without difficulty, with only *Nowaki* able to escape despite near misses by 16-inch shells. Numerous survivors were spotted but the Americans were proceeding at high speed and unable to assist the Japanese sailors. By the end of the action *Iowa* had fired 46 16-inch and 124 5-inch shells.

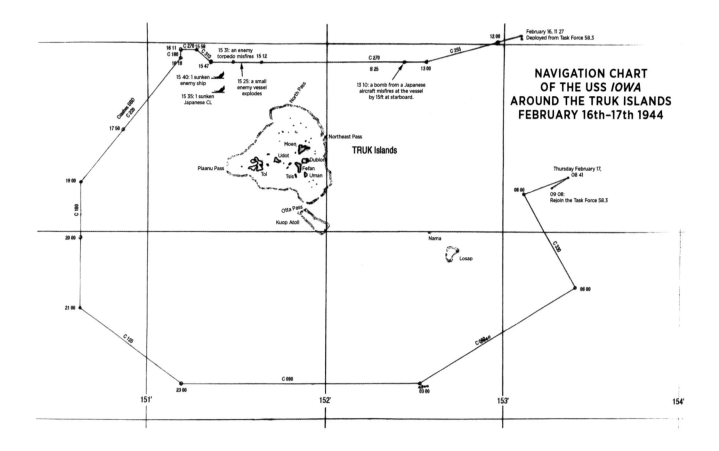

The Japanese light cruiser *Katori*.
USN

The destruction of the *Katori* under *Iowa*'s fire.
PRIVATE COLLECTION

One of *Iowa*'s 20-mm antiaircraft mounts in action.
NARA

Much as *Iowa*'s crew could be well satisfied with their ship, a technical problem had arisen in the closing of the breach of the center gun of turret 2 after the first salvo was fired that kept it from further participation in the action. The same was true of the left-hand gun of turret 3, where the breach was damaged by concussion. In addition, the two SG radars spent twelve minutes out of action as a result of defective cathode tubes. Despite these teething problems the ship had proved herself a remarkable engine of war.

Two days of raiding against and around Truk resulted in the destruction of thirty-two steamers, three light cruisers, six destroyers, and four auxiliaries for a total of 191,000 tons. In addition, two hundred aircraft were shot down and more than a hundred damaged. Unfortunately, the chaos reigning during those aerial battles resulted in a damaged Dauntless approaching the ship being shot down by *Iowa*'s 40-mm guns on the 16th. Five gun crews mistook the loudspeaker's "friendly plane" announcement for "enemy plane," with descriptions being changed to "own plane" and "bandit" during future air attacks.

On January 19, the squadron rejoined Task Group 58.3 and anchored at Kwajalein Atoll. By the end of the month the battleships were carrying out repeated training sorties with the carrier *Lexington* (CV 16), newly returned from repairs in the United States.

On March 17, after Rear Admiral Hustvedt had been replaced by Rear Adm. Willis Lee, *Iowa* and *New Jersey* weighed anchor and headed for Mili Atoll in the Marshall Islands. On the 18th each vessel took turns shelling the coastal defenses for ten minutes, ammunition depots and camouflaged installations being devastated by 16-inch shells at a range of 15,000 yards. The Japanese replied with 6-inch guns, *New Jersey* being bracketed by shells while *Iowa* suffered two hits on the port side. The first of these struck the armor of turret 2, destroying the optics of a telescopic sight and inflicting splinter damage on starboard no. 1 40-mm gun. The second shell, of 5-inch caliber, created a two-and-a-half by four-foot hole in the upper hull and deck at frame 134, but damage was negligible. Opening the range to 20,000 yards, *Iowa* fired 180 16-inch shells and *New Jersey* 187, while the 5-inch guns kept up continuous fire, which together silenced the coastal batteries.

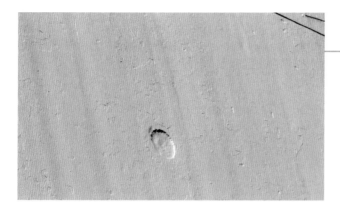

Iowa suffered two hits from a 6-inch and 5-inch coastal battery during the bombardment of Mili Atoll in the Marshall Islands. These photos show the impact on turret 2 in 1944 and as it appears today.
USN; PHILIPPE CARESSE

Turret 2 and the conning tower.
USN

Iowa and New Jersey returned to Majuro on the 19th. After dropping anchor Captain McCrea was surprised to receive a message from his counterpart in the Alabama (BB 60) inquiring as to the cause of the hole in his deck. "Rats," replied McCrea.

Reassigned to Task Force 58 at the end of that month, Iowa and New Jersey again found themselves supporting air raids, this time against the new Japanese airbase at Palau located on the western tip of the Caroline Islands. As shown in the table below, the raid on Palau (Operation Desecrate I) called for an impressive assemblage of naval power.

The new target was located more than two thousand miles west of Majuro. During its approach TF 58 was spotted by Japanese aircraft and had to repel an attack on the night of March 29. No damage was registered and New Jersey shot down two aircraft.

The major surface units went on to provide support during two days of air attack on Palau before shelling Woleai and its airstrip on April 1. This done, they turned for Majuro.

On the 13th, Task Force 58 sailed for Hollandia in New Guinea. Six days later it refueled north of the Admiralty Islands before shaping course for Wakde, which was reached on the 21st on the eve of the landings on Aitape led by Gen. Douglas MacArthur (Operation Persecution). The battleships were also present during the actions of Humboldt Bay and Tanahmerah Bay. On May 1

OPERATION DESECRATE I—TASK FORCE 58

BATTLESHIPS
North Carolina (BB 55)   South Dakota (BB 57)   Massachusetts (BB 59)
Alabama (BB 60)   Iowa (BB 61)   New Jersey (BB 62)

AIRCRAFT CARRIERS
Enterprise (CV 6)   Yorktown (CV 10)   Hornet (CV 12)
Lexington (CV 16)   Bunker Hill (CV 17)

ESCORT CARRIERS
Princeton (CVL 23)   Belleau Wood (CVL 24) Cowpens (CVL 25)
Monterey (CVL 26)   Langley (CVL 27)   Cabot (CVL 28)

HEAVY CRUISERS
Pensacola (CA 24)   Salt Lake City (CA 25) Louisville (CA 28)
New Orleans (CA 32)   Portland (CA 33)   Indianapolis (CA 35)
San Francisco (CA 38)   Wichita (CA 45)   Baltimore (CA 68)
Boston (CA 69)   Canberra (CA 70)

LIGHT CRUISERS
San Juan (CL 54)   Santa Fe (CL 60)   Biloxi (CL 80)

DESTROYERS
Dewey (DD 349)   MacDonough (DD 351)   Dale (DD 353)
Monaghan (DD 354)   Aylwin (DD 355)   Case (DD 370)
Hughes (DD 410)   Spence (DD 512)   Brown (DD 546)
Cowell (DD 547)   Charles Ausburne (DD 570) Charrette (DD 581)
Conner (DD 582)   Bell (DD 587)   Burns (DD 588)
Izard (DD 589)   Bancroft (DD 598)   Meade (DD 602)
Caldwell (DD 605)   Frazier (DD 607)   Edwards (DD 619)
Bullard (DD 660)   Callaghan (DD 792)   Charles P. Cecil (DD 835)

Iowa and *New Jersey* shelled the airfield at Ponape and caused heavy damage to the surrounding military installations. The fleet returned to Majuro on the 14th after an uneventful passage. This was followed in June by the start of Operation Forager aimed at conquering the Marianas. The fleet sailed on the 6th and a week later the battleships shelled Saipan and Tinian, with *Iowa* hitting an ammunition dump at a range of 10,000 yards, the detonation of which impressed those standing on the bridge. On the 11th and 12th the air wing of Task Group 58.4, of which *Iowa* and *New Jersey* were part, accounted for twenty-two merchantmen. After refueling at sea 130 miles northeast of Saipan on the 14th, TG 58.4 fell in with TF 58 three days later and steered for the Philippine Sea.

At 0550 on the 19th, the U.S. fleet was spotted by a Zero, which managed to report its position before being shot down. A violent aerial battle ensued during which *Iowa* was repeatedly engaged by Japanese aircraft. An attack by a Kate torpedo bomber was narrowly avoided and the assailant was shot down together with two others. At day's end the battleships turned to refuel at Guam.

On the 21st, the battleships and their escort were detached to pursue the fleet led by Admiral Ozawa Jisaburō, which consisted of the carriers *Zuikaku*, *Zuiho*, *Chitose*, and *Chiyoda*, the hybrid battleships *Hyuga* and *Ise*, the heavy cruisers *Oyodo*, *Isuzu*, and *Tama*, and eight destroyers. No contact was made and *Iowa* and *New Jersey* rejoined the main fleet on the 24th.

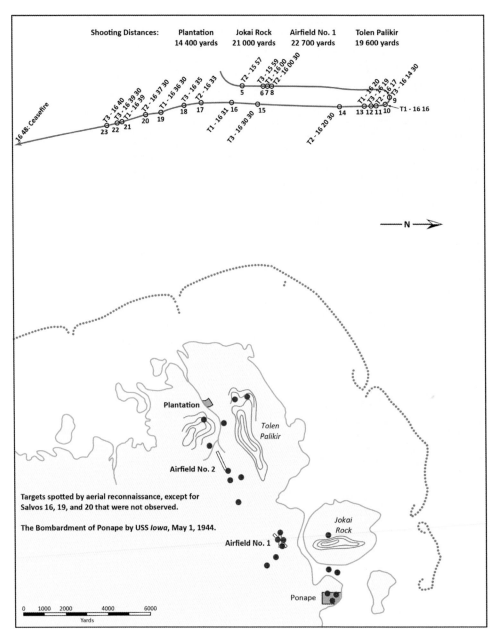

Task Force 58 anchored at Majuro. *New Jersey* is in the center of the photo with *Iowa* wearing her distinctive camouflage on the right. Beyond her are *Washington* (right) and *North Carolina*.
NARA

This photo taken in 1985 illustrates the violence of a 16-inch salvo. *Iowa* carried out numerous bombardments of Japanese positions during the Pacific War. NARA

A successful bombardment of Pagan was carried out on the 28th and twenty-four hours later it was Rota's turn to come under 16-inch shellfire. On July 4, Task Group 58.4 steered for Eniwetok, where it anchored on the 6th. After a brief respite the fleet sailed for a waiting position seventy miles southeast of Guam, which was bombarded on the 17th, 18th, and 23rd. Tinian suffered the same punishment on the 26th, after which the fleet turned for rearming and replenishment at Saipan while *New Jersey* made her way to Pearl Harbor. During this prolonged campaign Captain McCrea had spent sixty-eight days and nights on his bridge without once entering his cabin. It was the same in the conning tower for Rear Admiral Hustvedt, who only visited his stateroom to shower. Task Group 58.4 reached Eniwetok on August 9 and later that month Captain McCann took command of *Iowa*.

On September 6, Adm. William F. Halsey, commander in chief of the Third Fleet and then flying his flag in *New Jersey*, joined Task Force 38 including *Iowa* as part of Task Group 38.2. Two days later the battleships refueled from the oilers *Lackawanna* (AO 40) and *Cimarron* (AO 22) before escorting the carriers on raids against Luzon. After a stop at Garapan Harbor on Saipan on the 28th, they sailed for Ulithi in the Caroline Islands the following day. On October 12, Task Group 38.2 bombarded the airfields of Shinihi and Macujama on Formosa, after which *Iowa* and *New Jersey* made preparations to join Operation King II, the invasion and seizure of Leyte in the Japanese-occupied Philippines.

These plans, however, were forestalled by the launch by the Japanese of Operation Shō-Ichi-Gō ("Victory") on October 20, heralding the start of one of the greatest air-naval engagements in history and fought over an immense area. The plans prepared by Admiral Toyoda Soemu involved luring the U.S. aircraft carriers north in order to expose the transports to attack by Japanese battleships and cruisers. Meanwhile, the bulk of the Japanese fleet commanded by Admiral Kurita Takeo was to intercept and destroy the U.S. invasion forces after a night passage through the San Bernardino Strait. Things did not go to plan, however, and Kurita had already lost three heavy cruisers to submarine attack by the 23rd.

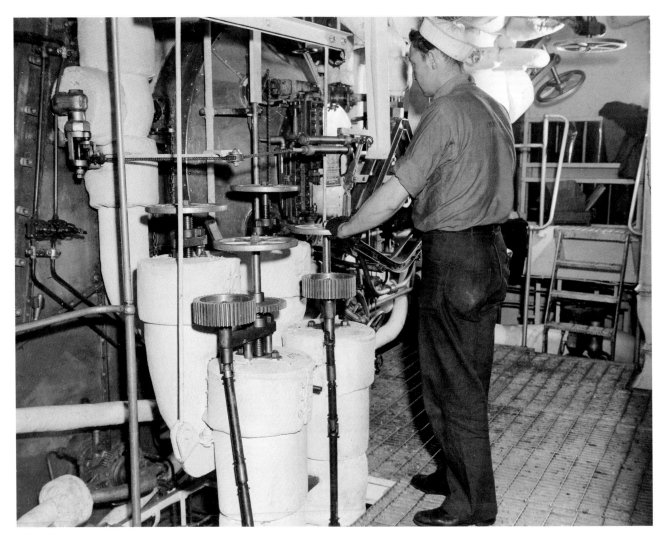

Iowa's firemen and machinists had their work cut out keeping the ship in good operating order. This photo shows one of the firerooms.
NARA

Units of Admiral Kurita's Second Fleet shortly before the start of Operation Shō-Ichi-Gō, which resulted in the Battle of Leyte Gulf. From right to left are the battleship *Nagato*, a heavy cruiser, and the battleships *Yamato* and *Musashi*.
USN

Iowa and the carrier *Cabot* (CVL 28) during the Battle of Luzon on November 25, 1944. USN

The following day the heavy cruiser *Myoko* was so heavily damaged that she was required to return to Brunei, and at 1936 the giant battleship *Musashi* succumbed to twenty torpedoes and eighteen bombs from U.S. aircraft. Faced by these losses, Kurita turned back to regain the open water while Halsey, still flying his flag in *New Jersey*, looked to take advantage of a weakened enemy. He therefore set off in pursuit of Kurita at 28 knots with *New Jersey*, the light cruisers *Vincennes* (CL 64), *Miami* (CL 89), and *Oakland* (CL 95), and eight destroyers. Halsey arrived three hours too late, leaving the outcome of an engagement between Kurita's flagship *Yamato* and his own a matter for conjecture. With a full-load displacement of 75,400 tons and nine 18.1-inch guns, *Yamato* and her sisters were the only battleships capable of meeting the *Iowa* class on an equal footing. Whereas the *Iowa*s had superior fire control and radar and a high-capacity armor-piercing shell, *Yamato* fired salvos of 28,968 pounds and enjoyed a sophisticated scheme of protection.

Later that day *Iowa* destroyed a Yokosuka D4Y Judy dive bomber with the expenditure of 108 rounds of 40-mm and twenty-three rounds of 20-mm ammunition. Attacks were launched on Manila Bay on the 28th and 29th during which *Iowa* faced her first kamikaze attack. A series of actions in support of operations by Task Group 38.2 kept *Iowa* from Ulithi anchorage until November 9. Five days later she sailed to Philippine waters to relieve Task Group 38.3, which consisted of the carriers *Lexington* (CV 16), *Essex*, *Ticonderoga*, *Langley* (CVL 27), *Princeton*, the battleships *Massachusetts*, *Alabama*, *Indiana*, *South Dakota*, four light cruisers, and sixteen destroyers.

Launching a Vought OS2U Kingfisher floatplane on October 26, 1944.
USN

Iowa at speed in the Pacific.
NARA

Iowa entering floating dry dock *ABSD-2*.
USN

On the 25th, the Third Fleet faced repeated attacks from kamikaze aircraft during which *Iowa* destroyed two Nakajima B6N Tenzan Jill torpedo bombers and a Judy reconnaissance aircraft that came down just a hundred yards from her. The carriers *Intrepid* (CV 11), *Hancock* (CV 19), and *Cabot* (CVL 28), however, were not so fortunate, suffering extensive damage from suicide attacks.

On December 18, the fleet, recently reinforced by *Wisconsin*, was assailed by Typhoon Cobra (known as "Halsey's Typhoon"), which accounted for the destroyers *Hull* (DD 350), *Monaghan* (DD 354), and *Spence* (DD 512) with heavy loss of life, while a cruiser, five aircraft carriers, and three destroyers were severely damaged. Around 790 men were lost with more than 80 injured. The storm struck during a refueling operation, and it was necessary to await a lull in order to resume refueling, with the fleet proceeding at low speed three hundred miles east of Luzon.

Iowa and *New Jersey* each had a floatplane lost overboard on the 18th, but five days later severe vibration was felt on propeller shaft 3 as *Iowa* was increasing speed from 18 knots. The propeller had suffered damage, which restricted speed to 15 knots and required an inspection of the hull at Manus in the Admiralty Islands. With the heavy cruiser *Canberra* and the destroyers *Claxton* and *Killen* refitting in floating dock *ABSD-2* at Manus, Admiral Halsey informed its commander, Capt. R. K. James, that he had seventy-two hours to clear the dock in readiness to receive *Iowa*. Waiting on board a tug to bring *Iowa* into the dock, James was alarmed when *Iowa* declined assistance and entered the dock directly at 5 knots, the navigating officer ordering full astern and stopping the ship in the desired position.

With Manus lacking the facilities to carry out full repairs, *Iowa* sailed for Hunters Point Shipyard in San Francisco, where she was docked from January 15 to March 19, 1945. During this time the opportunity was taken to enclose the navigating bridge and install large bay windows.

Iowa refitting at Manus on December 28, 1944.
USN

Iowa at Hunters Point on March 18, 1945.
USN

Divine service on board *Iowa* in 1945 as seen from the escort destroyer USS *Cabana* (DE 260).
USN

The oiler *Cahaba* (AO 82) refueling *Iowa* and the carrier *Shangri-La* (CV 38) on July 8, 1945.
USN

This completed, *Iowa* sailed for Okinawa, which she reached on April 15 to relieve *New Jersey*, with Rear Adm. Oscar C. Badger shifting his flag to her on that date. She was then assigned to Task Group 38.4 together with *Missouri*, the carriers *Yorktown* (CV 10), *Ticonderoga* (CV 14), *Shangri-La* (CV 38), the large cruisers *Alaska* (CB 01) and *Guam* (CB 02), two light cruisers, and nine destroyers. This force was strengthened by the addition of *Wisconsin* in June.

On June 10, the battleships bombarded Minami Ogari Shima and Minami Daito Shima east of Okinawa, after which they anchored at San Pedro Bay off Leyte on the 13th. Here they remained until July 1, *Iowa* having by then come under the command of Capt. Charles Wellborn. On the 15th, a bombardment of industrial installations on Hokkaidō was carried out as well as against Muroran, during which the three battleships fired a total of 833 16-inch shells. Forty-eight hours later *Iowa* hoisted the flag of Rear Admiral Badger, who ordered steam for 24 knots in order to shell the industrial center of Hitachi Miro on Honshū, during which many of the ship's radar aerials were damaged by blast. On the 29th and 30th it was the island of Kahoʻolawe's turn to be bombarded while aircraft hit targets around Tokyo. Further shelling was carried out against Honshū and Hokkaidō on the 8th, with *Iowa* firing 697 16-inch shells in fifty-one minutes.

Iowa followed by the Japanese destroyer *Hatsuzakura* carrying a pilot to allow *Missouri* to enter Tokyo Bay in safety.
NARA

Iowa in 1945.
NARA

The intensity of the bombardments significantly weakened Japanese industrial capacity as the conflict drew to a close, and on August 14 the Imperial Council issued a declaration accepting unconditional surrender following the attacks on Hiroshima and Nagasaki.

On the 27th, *Iowa* and *Missouri* joined twenty-three carriers, twelve heavy cruisers, twenty-six light cruisers, twelve submarines, and 185 other vessels in the surrender of the Yokosuka naval base. Two days later she was in Tokyo Bay supporting invasion forces and was present at the historic surrender of the Japanese Empire on September 2. *Iowa* quit Japanese waters on September 20. Before returning home *Iowa* participated in Operation Magic Carpet, the repatriation of large numbers of troops from the Pacific War. She reached Seattle on October 15 before sailing for Long Beach Naval Base, her home port.

After a spell of refitting and exercises, *Iowa* returned to Tokyo Bay on January 27, 1946, where she became flagship of the Fifth Fleet and entered her peacetime routine. On March 25 she again pointed her bows toward Long Beach, which she reached after stops at several West Coast ports punctuated by exercises en route. By October new radar had been fitted and many 20-mm and 40-mm mounts removed. In late July *Iowa* carried out target practice on the decommissioned battleship *Nevada* (BB 36), already a survivor of the Bikini atom bomb tests. This was followed by a series of training cruises including a visit to Vancouver before embarking the governor of California for transport to Hawaii in July 1947. There were few sorties the following year, which *Iowa* spent largely at Long Beach and Mare Island. By September 1948 she was in San Francisco, being decommissioned for the first time on March 24, 1949.

Iowa's stern in Puget Sound on October 11, 1946.
NARA

The forecastle seen in 1946.
USN

Iowa in San Francisco Bay in 1946.
USN

Iowa in 1947.
KARA

Iowa shortly before her first decommissioning.
J. PRADIGNAC

Iowa's stern in Puget Sound on October 11, 1946.
NARA

The forecastle seen in 1946.
USN

Iowa in San Francisco Bay in 1946.
USN

Iowa in 1947.

Iowa shortly before her first decommissioning.
J. PRADIGNAC

Iowa in 1947.
NARA

The Korean War

On June 25, 1950, North Korea invaded South Korean territory. The failure by the United Nations Security Council to impose a cease-fire resulted in a decision by UN member states to provide military support to the south.

The capacity of the *Iowa*-class ships to deliver fire support to ground troops immediately made them candidates for reactivation, although *Iowa* herself was the last of the four to be brought out of mothballs at Suisun Bay near San Francisco. Taking command on August 25, 1951, Capt. William Smedberg found the vessel in relatively good condition. A crew was drafted for him in under three weeks, but the main problem was finding men competent to operate the 16-inch guns. Smedberg began running speed trials at 20 knots near Alcatraz and the Golden Gate, but the strong currents made maneuvers problematic and there was a severe threat of grounding. Adm. Chester W. Nimitz, then in retirement, witnessed these trials and, acutely aware of the danger since he himself had run a destroyer aground in 1908, lectured Smedberg on the subject. Adm. Clifton A. Sprague, who was kept abreast of these conversations, ordered *Iowa* transferred to San Diego. The access to that port, however, was no more than thirty-six feet deep and there was a significant risk of her taking ground there too. The battleship was therefore anchored off Coronado, a decision heavily criticized by both admirals, and Smedberg was relieved of command in July 1952.

On April 8 *Iowa* fired her first shells against Chahochŏn and two days later destroyed a series of fortified positions, with an enemy shell exploding 250 yards off the ship's starboard beam. After refueling at sea, on the 11th she opened fire on targets in support of the US X Corps at a range of 32,800 yards. *Iowa* soon developed problems with her boilers, however, which reduced her best speed to 27 knots except in emergencies. Nonetheless, on the 22nd the commander in chief of the Pacific Region (CINCPAC), Adm. Arthur W. Radford, flew by helicopter to *Iowa* to supervise her bombardment operations. On the 25th she destroyed the village of Chindong and by the end of that month had shelled enemy positions at Wonson, Konsong, and Tanchon, where she destroyed a railway line and accounted for a hundred North Korean soldiers. Over the course of nine days she fired 549 16-inch shells and 1,486 5-inch shells.

Next she bombarded a locomotive depot and several industrial sites before shelling the North Korean port of Chongjin on May 25 while escorted by four destroyers, with air cover provided by two F4U Corsairs and the ship's helicopter. At around 0500 *Iowa*'s air-warning radar detected four unknown aircraft near the Manchurian frontier. These were soon joined by another ten unidentified aircraft, and tension began to mount on board *Iowa*, but no sooner had four F9F Panther fighters appeared than the radar contacts vanished. The bombardment could then be carried out without difficulty, and at noon Radio Peking declared that U.S. ships had deliberately shelled civilian positions. Three hours later it announced the sinking of three U.S. vessels although the squadron was in fact undamaged. On May 28 the ship was again supporting X Corps and, as before, inflicted considerable destruction on enemy positions.

Iowa leaving the Reserve Fleet on July 14, 1951.
NARA

Iowa's recommissioning ceremony on August 25, 1951.
NARA

Iowa in a floating dock at Hunters Point Shipyard in 1952.
USN

Iowa seen in 1952 during the Korean War.
USN

Iowa shelling the Korean coast on May 27, 1952.
NARA

On June 1 she turned for Sasebo for replenishment, but was soon back in action, and one of her helicopters rescued a pilot from the *Princeton* (CV 37) on the 9th. The following day *Iowa* appeared off Mayang Do, where she brought down a key bridge. A few days later she was assisting *LST 692* en route to Yodo Island, and on the 16th she rescued another of *Princeton*'s pilots, this time thirty-five miles northwest of Hungnam. That same month Captain Smedberg was replaced by Capt. Joshua Cooper, a keen advocate of the helicopter, who lost no opportunity to go aloft in one, by day or by night and in all weather. The attractions of this aircraft, however, lessened greatly after an incident in which the pilot failed to locate the ship in heavy fog and only chanced upon her by miracle.

On August 20 *Iowa* was ordered to embark numerous injured sailors from the destroyer *Thompson* (DD 627), which had been taken under fire by a concealed Chinese battery near Songjin. A shell had exploded on the bridge, killing four crewmen and wounding nine. At her best speed, she immediately made for *Iowa*, which lay sixteen miles to the south and had adequate medical facilities on board.

On September 23 *Iowa* was visited by Gen. Mark Clark, commander in chief of UN forces and ambassador to South Korea, to oversee the bombardment of Wonsan during which numerous ammunition depots were destroyed.

On the 25th, in company with the British destroyer HMS *Charity*, she fired forty-five 16-inch shells, accounting for a locomotive and thirty wagons and other equipment. Steaming off Tanchon, she was able to pick up a number of refugees who had put to sea in a small boat under the white flag.

Before returning to the United States, *Iowa* participated in Operation Decoy, which involved an infantry landing in the Kojo area to draw an enemy response, in connection with which she and the heavy cruiser *Toledo* (CA 133) provided gunfire support on October 14. During this operation an aircraft from the carrier *Bonhomme Richard* (CV 31) was obliged to ditch, but *Iowa*'s helicopter was on hand to rescue the pilot. On the 16th *Iowa* provided antiaircraft support to the amphibious command ship *Mount McKinley* (AGC 7). On October 19 *Iowa* sailed from Yokosuka for Norfolk via Pearl Harbor and decommissioning. She had steamed over 40,000 miles during the Korean War.

After a spell of training in the Caribbean, in July 1953 *Iowa* joined NATO forces in Operation Mariner for which Vice Adm. E. T. Woolfidge, commander in chief of the Second Fleet, hoisted his flag on board. This exercise was carried out in the Atlantic and the North Sea under the ship's new commander, Capt. Wayne Loud.

On June 7, 1954, cadets embarked for training purposes had the good fortune to be present on the only occasion

The carrier *Philippine Sea* (CV 47), destroyer *Barton* (DD 722), and *Iowa* operating off Korea.
USN

Iowa in 1952.
J. PRADIGNAC

Iowa followed by *New Jersey* shortly after the Korean War.
USN

Iowa lying off Taarbæk near Copenhagen.
PHILIPPE CARESSE

on which the entire *Iowa* class (2nd Battleship Division) sailed in company, the four ships maneuvering off the Virginia Capes.

In January 1955, *Iowa* sailed for the Mediterranean, where she visited Gibraltar, Mers el Kébir, Genoa, Naples, Istanbul, Athens, and Cannes. On her return the ship made stops at Barcelona and Portsmouth in England before returning to Norfolk in August for refitting. In December the 16-inch gun barrels were replaced, having fired 8,279 shells of all types since the ship had entered service. After the inevitable steam trials, *Iowa* visited Havana between April 13 and 15, 1956. The cadet cruise that followed in June took them to Guantánamo Bay, Bermuda, Portsmouth, and Copenhagen.

In January 1957 *Iowa* was back in the Mediterranean with the Sixth Fleet before participating in the International Naval Review held at Hampton Roads, Virginia, between June 8 and 17. Three hundred ships from sixteen nations commemorated the 350th anniversary of the founding of Jamestown, Virginia.

The forward superstructure in 1955.
USN

All four units of the *Iowa* class maneuvering off the Virginia Capes on June 7, 1954, the only occasion on which all four units sailed in company.
USN

A Piasecki HUP seen on *Iowa*'s turret 1, the ship en route to Cuba.
USN

An immense flypast took place at 1520 on the 12th, after which the ships cruised off Fort Wool, Willoughby Spit, and Old Point Comfort. This was followed by a Marine landing exercise on the 14th before the ships quit the anchorage on the 17th. Between September 3 and 12, *Iowa* and *Wisconsin* steamed to Scotland to join NATO units in Operation Strikeback, aimed at countering a possible Soviet attack on Europe. The operation involved numerous vessels of the U.S., Canadian, British, French, Dutch, and Norwegian navies.

These exercises were carried out in the Atlantic, the Norwegian Sea, and the GIUK (Greenland, Iceland, United Kingdom) area and involved nine aircraft carriers, two battleships, eight cruisers, fifty-one destroyers, twenty-five submarines, and five auxiliaries. *Iowa* returned to Norfolk on September 28 and sailed for Hampton Roads on October 22. She decommissioned for the third time in Philadelphia on February 24, 1958.

Iowa en route to be decommissioned for the second time on October 23, 1957.
USN

Iowa decommissioned in Philadelphia.
NARA

Iowa, *Wisconsin*, and the carrier *Shangri-La* (CV 38) decommissioned in Philadelphia in 1981.
USN

The Function of a Battleship in Modern Naval Warfare

Iowa did not see action during the Vietnam War, and once it was over few sailors believed the *Iowa*-class battleships would ever put to sea again. But the defense policy of President Ronald Reagan and Secretary of the Navy John Lehman's determination to have a six-hundred-ship Navy meant that it was not long before the battleships returned to service, a decision assisted by the commissioning of the 28,000-ton Soviet guided-missile cruiser *Kirov* in December 1980. Aircraft carriers aside, *Kirov* was then the most powerful warship afloat. The U.S. Navy had no comparable construction program so it was decided to return the *Iowa* class to service. Many conversion programs were studied, notably that of removing Turret 3 to permit operations by the AV-8B Harrier II vertical launch aircraft or the installation of missile launchers. These proposals were not pursued and were eventually shelved in 1984. Each vessel, however, was to receive modern electronic systems and the latest antiaircraft defense together with missiles. The cost of modernizing each vessel was in excess of $400 million.

The *Kirov*-class guided-missile cruiser *Pietr Velykiy*.
PRADIGNAC AND LEO

Iowa under tow on April 22, 1982, on her way to be modernized.
INGALLS SHIPBUILDING

Iowa being towed to the Gulf of Mexico for modernization.
J. PRADIGNAC

The first battleship to be recommissioned was *New Jersey*, followed soon after by *Iowa*, for which funds were voted in the 1982 budget. On September 1 that year the latter was towed by the tugs *Apache* (ATF 172), *Paiute* (ATF 159), and *Pagano* (ATF 160) to New Orleans for refitting, arriving on the 16th. She was then ordered to Ingalls Shipbuilding of Pascagoula, Mississippi, for completion. The challenges of restoring the dilapidated condition of the ship after twenty-six years of immobilization are to be imagined. *Iowa* largely preserved her silhouette, however, despite the addition of new detection, communication, and weapons systems. Visible in the superstructure was her enhanced weapons fit of thirty-two BGM-109 Tomahawk and sixteen RGM-84 Harpoon missiles while four of the after twin 5-inch mounts were removed.

Iowa recommissioned under Capt. Gerald Gneckow on April 28, 1984, and was operational a year ahead of schedule, although the Board of Inspection and Survey (InSurv), which had responsibility for inspecting and assessing the material condition of U.S. Navy vessels, did not survey the ship's engineering and gunnery systems. The recommissioning ceremony was attended by Vice President George H. W. Bush and Secretary of the Navy John Lehman. A guest of honor among the attendees was none other than Vice Adm. John L. McCrea, her first commanding officer.

Iowa being refitted at Ingalls Shipbuilding prior to returning to service, September 23, 1983.

Iowa shortly before being recommissioned into the U.S. Navy.
NARA

Vice President George H. W. Bush delivers a speech during *Iowa*'s recommissioning ceremony on April 28, 1984.
NARA

Main battery exercises off Vieques Island on July 1, 1984.
USN

Thus refitted, *Iowa* headed for Guantánamo Bay on her shakedown cruise. She was still capable of speeds in excess of 32 knots. By now rechristened "The Big Stick," her first mission was to be in support of the U.S. naval presence off Lebanon, but this had been significantly reduced by policy changes by the time of her appearance.

Iowa spent two weeks at Guantánamo Bay in May 1984 and the 21st found her off Vieques Island for the calibration of her 16-inch and 5-inch guns. On June 19, *Iowa* sailed for Caracas with the intention of carrying out gunnery exercises en route. On August 12 *Iowa* launched two Army UH-1 helicopters to provide medical assistance in Guatemala, after which the battleship received Guatemalan dignitaries on board, providing tours and giving a firepower demonstration before conducting maritime reconnaissance thirteen miles off the Nicaraguan coast. On the 26th she reached Balboa to traverse the Panama Canal, making her home port of Norfolk on September 17 before spending a week in New York in October.

Iowa carries out a 16-inch shoot on August 15, 1984.
USN

Iowa in the Pedro Miguel lock of the Panama Canal in August 1984.
NARA

Iowa in one of the locks of the Panama Canal in August 1984.
NARA

Iowa traversing the Panama Canal.
USN

Iowa at sea in 1984.
NARA

Iowa docked at Norfolk in the spring of 1985.
USN

In February 1985 *Iowa* provided humanitarian assistance to Costa Rica and Honduras, after which she was drydocked at Norfolk until April 26. Emerging from this on July 31 she rearmed at Hampton Roads before carrying out Tomahawk firing exercises off Yorktown, Virginia.

In August 1985 the NATO Ocean Safari '85 exercise assembled 167 warships, including *Iowa*, together with Marine contingents from Belgium, Canada, Denmark, France, the Netherlands, Norway, Portugal, the United Kingdom, and West Germany. This exercise took *Iowa* and her battle group across the Arctic Circle in what were often immense seas as a demonstration of force against the Soviet naval threat. On the conclusion of the exercise in September, *Iowa* joined the carrier *America* (CVA 66) on the 20th for stops at Le Havre, Copenhagen, and Oslo.

From October 12 she participated in Operation Baltops '85, which brought her into the Baltic Sea, and she used the opportunity to visit the German port of Kiel. Baltops was organized to improve the U.S. Navy's tactical control over a key strategic naval and air environment at the gates of what was then largely a Soviet lake. Six U.S. vessels participated including *Iowa*, *Ticonderoga* (CG 47), *Aylwin* (FF 1081), *Halyburton* (FFG 40), *Pharris* (FF 1094), and *Merrimack* (AO 179). Sea rescue, firing exercises, mine warfare, underway replenishment, and antisubmarine exercises were successfully carried out. On the 26th *Iowa* set a course for Norfolk while continuing crew training. The vessel ended the year operating off Central America.

Iowa docked at Norfolk in April or May 1985.
USN

Iowa during the NATO Ocean Safari '85 exercise.
NARA

Iowa tied up near Akershus Castle, Denmark, on October 11, 1985.
NARA

Iowa under way in Oslo fjord on October 26, 1985.
NARA

Iowa at the French port of Le Havre on September 23, 1985.
PRIVATE COLLECTION

Heavy shells awaiting stowage in the magazines in 1985.
USN

Now under the command of Capt. Larry Seaquist, on July 4 *Iowa* embarked President Reagan, First Lady Nancy Reagan, and Secretary of the Navy John Lehman to participate in a major naval review in the Hudson to commemorate the centennial of the Statue of Liberty. Three days of festivities began with the firing of salutes by vessels from thirty-five nations for "Liberty Weekend."

President Ronald Reagan and First Lady Nancy Reagan on board *Iowa* on July 4, 1986.
NARA

Iowa in New York to celebrate the centenary of the Statue of Liberty.
NARA

The following months were taken up with exercises and training off the Florida coast and in the Gulf of Mexico. On August 17 *Iowa* sailed for the North Atlantic to join NATO units as part of Operation Northern Wedding, during which 150 ships from ten countries steamed toward southern Norway, where landing operations were successfully carried out. The ships spent twenty-four days at sea without interruption. Sadly, a CH-46 Sea Knight helicopter from *Saipan* (LHA 2) ditched west of Bodø at 1130 on August 29 with twenty-one occupants, of whom eight were lost. Between September 5 and 6 *Iowa* carried out gunnery exercises off Cape Wrath, Scotland, firing nineteen 16-inch and thirty-two 5-inch shells. At the end of that month *Iowa* stopped at Portsmouth, England, and Bremerhaven, Germany, where she was open to the public before turning for Norfolk on October 2.

On November 9 *Iowa* served as base ship off the Virginia Capes during the commissioning trials of the RQ-2B Pioneer drone, and on the 14th she fired her thousandth 16-inch shell since returning to active service in 1984.

On January 9, 1987, *Iowa* left Norfolk for the Caribbean to participate in the BLASTEX 1-87 Exercise, stopping at Honduras, Colombia, and the Virgin Islands. In February she headed for Guantánamo Bay and again carried out firing exercises off Vieques. After a spell at Norfolk, on May 3 she participated in Operation SACEX (Supporting Arms Coordination Exercise) consisting of supporting an amphibious assault in the Greater Antilles, particularly Puerto Rico. Between July 8 and 26 *Iowa* participated in FLEETEX 3-87 in the Atlantic before putting in at Yorktown for rearming. On September 10 she sailed from Norfolk for the Mediterranean, where she joined the Sixth Fleet on the 20th. Two days later she joined Exercise Display Determination, stopping at Istanbul on October 8. On the 22nd she was detached from the Sixth Fleet and reached the Norwegian port of Trondheim on the 30th. She was soon back in the Mediterranean, however, and on November 25 traversed the Suez Canal as part of Operation Earnest Will following a Kuwaiti appeal to the Reagan administration to protect maritime trade passing the Strait of Hormuz. Fifteen vessels of the U.S. Navy were involved in this operation to protect oil tanker convoys, and *Iowa* reached Diego Garcia in the Chagos Archipelago on December 4 prior to starting patrol work in the Persian Gulf. *Iowa* and her battle group were responsible for escorting tankers between the Kuwaiti port of Mina Al-Ahmadi and the Strait of Hormuz.

After lengthy patrols in the Indian Ocean, on February 29, 1988, *Iowa* turned for Norfolk, again traversing the Suez Canal and reaching her home port on March 10, where the ship was laid up for refitting and the crew granted leave. On April 20, *Iowa* sailed for New York to attend the traditional Navy Week celebrations held between the 21st and 25th, during which eight warships were open to the public. Having rearmed at Yorktown, *Iowa* found herself under a new commanding officer on May 23, Capt. Fred Moosally.

Preparing to launch an RQ-2B Pioneer drone.
NARA

Iowa in Norwegian waters on September 1, 1986.
NARA

Iowa carrying out speed trials in Chesapeake Bay.
NARA

Iowa under way during Operation Northern Wedding.
NARA

Iowa fires a Harpoon RGM-84 antiship missile.
USN

Iowa firing a Tomahawk BGM-109 cruise missile on August 2, 1986.
USN

Turrets 1 and 2 in action.
USN

Iowa refueling from the carrier USS *Midway* (CV 41).
NARA

Iowa sailing in company with the West German guided-missile destroyer *Schleswig-Holstein* (D 182).
NARA

Iowa sailing in close order with the Danish frigate *Peder Skram* (F 352) (*left*) and the German guided-missile destroyer *Rommel* (D 187) (*right*).
NARA

The Explosion in Turret 2

On April 19, 1989, an explosion occurred in *Iowa*'s turret 2 while the ship was conducting gunnery exercises off Vieques. As the turret crew prepared to fire, the powder bags from the center gun ignited before the breech was closed. The resultant deflagration and fire spread throughout the turret structure as far as the powder and projectile flats several decks below.

Forty-seven sailors lost their lives in turret 2, but there were eleven in the magazines surrounding the turret who survived without serious injury. These crewmen were separated from the interior powder-handling flat by an annular space designed to isolate the ship's magazines from an explosive flash, a feature that allowed them to escape from the environment above.

Iowa's damage-control crews responded quickly. Their efforts in fighting the resultant fires prevented the remaining powder and projectiles inside the turret from igniting. The magazines were flooded and the fires extinguished in about ninety minutes through heroic efforts by the crew, thereby preventing additional loss of life and possibly saving *Iowa* herself.

Capt. Fred Moosally.
NARA

ACCIDENT OR INTENTIONAL ACT?

While the Navy technical team began searching for a purely mechanical reason for the accident, Navy investigators discovered a Department of Defense explosives manual together with a volume titled *Getting Even: The Complete Book of Dirty Tricks* inside the personal locker of one of the gun captains who perished in the incident. This, coupled with a suspicious life insurance policy taken out by the same individual along with unsubstantiated allegations of a failed homosexual affair, led some investigators to believe that the accident may have been an intentional act on the part of the gun captain. This theory gained momentum when the technical team could find no evidence of mechanical failure as a cause of the accident. The technical team therefore began looking for evidence of an explosive device that may have been used to initiate the turret explosion.

THE SEARCH FOR EVIDENCE OF AN EXPLOSIVE DEVICE

The Navy determined that evidence of an explosive device would need to be found somewhere in the vicinity of the source of the explosion in order to confirm the veracity of the intentional act theory. Since the explosion would have consumed most of any such device, the Navy theorized that chemical traces of this device might have been trapped in the compressed cannelure or fin of the rotating band on the projectile in question when the explosion occurred. Materials analysis of cannelure specimens by Navy laboratories found chemical indicators they claimed could only have been produced by such a device, and they used these conclusions to support the intentional act theory. Later analysis by FBI and Sandia National Laboratories disagreed with this conclusion and claimed that the presence of an initiating explosive device could not be "proved or disproved" based on the available evidence.

PROPELLANT SENSITIVITY

During the initial technical investigation, the Navy had tested the propellant for sensitivity to ignition by impact, friction, heat, or electrical sources and found it to be stable under all these conditions. This, coupled with more than fifty years of experience with similar gun systems, led the Navy to conclude that propellant sensitivity was not a causal factor in the accident. In response to pressure from the families of the victims and public outcry over the conclusions of the initial investigation, however, Congress instructed the General Accounting Office to commission Sandia National Laboratories to carry out an independent investigation into the Navy's conclusions.

During its investigation into propellant stability, Sandia discovered that the propellant had a previously unknown sensitivity to impact under conditions that might occur during the loading of the guns on board a battleship. Specifically, it found that the propellant pellets in a powder bag might ignite if the tare or trim layer used to adjust the overall weight of each bag to specification by the addition or subtraction of individual pellets were reduced in such a way as to make them liable to fracture more easily under the impact of a potential overram situation, in which the bag was pushed too far into the chamber of the gun. This conclusion represents a plausible alternative theory that might explain the accidental ignition of the propellant during the incident.

OVERRAMMING

The Navy had concluded early in its technical investigation that the powder had been overrammed, but they believed this to have no bearing on the ignition of the propellant. Sandia, however, came to a different conclusion. Its analysis of the overram concluded that the powder or propellant had been pushed so far into the chamber that it had impacted against the base of the projectile, thereby creating sufficient force to fracture one or more of the propellant pellets and potentially cause ignition. Sandia based this conclusion on its analysis of gouges in the spanning tray they believed demonstrated that the ramming device had been deeper inside the chamber than the Navy had calculated. The Navy disagreed with Sandia's conclusions and provided additional testing in support of its position. The actual depth of the overram was never conclusively determined.

INCONCLUSIVE EVIDENCE

While there are several plausible explanations as to why the propellant in *Iowa*'s gun ignited, lack of evidence has made it impossible to reach a definitive conclusion.

Iowa's turret 2 was never fully repaired and remains sealed. A memorial to the forty-seven sailors killed in the explosion was erected at the Norfolk Naval Station, *Iowa*'s last home port, and the spot was renamed Iowa Point. Each year on April 19 a memorial service is held there, as well as on board *Iowa* herself at her permanent berth at San Pedro, California.

A propellant ignition claims the lives of forty-seven sailors in turret 2 as the ship steams 330 miles northeast of Puerto Rico. USN

Crewmen battle to extinguish the fires raging in turret 2 on the morning of April 19, 1989. USN

Iowa returning to Norfolk with her immobilized turret 2 still trained to starboard.
NARA

THE VICTIMS OF THE EXPLOSION IN *IOWA*'S TURRET 2, APRIL 19, 1989

Tung Thanh Adams, 25, Fire Controlman 3rd Class
Robert Wallace Backherms, 30, Gunner's Mate 3rd Class
Dwayne Collier Battle, 21, Electrician's Mate, Fireman Apprentice
Walter Scot Blakey, 20, Gunner's Mate 3rd Class
Peter Edward Bopp, 21, Gunner's Mate 3rd Class
Ramon Jarel Bradshaw, 19, Seaman Recruit
Phillip Edward Buch, 24, Lieutenant Junior Grade
Eric Ellis Casey, 21, Seaman Apprentice
John Peter Cramer, 28, Gunner's Mate 2nd Class
Milton Francis Devaul Jr., 21, Gunner's Mate 3rd Class
Leslie Allen Everhart Jr., 31, Seaman Apprentice
Gary John Fisk, 24, Boatswain's Mate 2nd Class
Tyrone Dwayne Foley, 27, Seaman
Robert James Gedeon III, 22, Seaman Apprentice
Brian Wayne Gendron, 20, Seaman Apprentice
John Leonard Goins, 20, Seaman Recruit
David L. Hanson, 23, Electrician's Mate 3rd Class
Ernest Edward Hanyecz, 27, Gunner's Mate 1st Class
Clayton Michael Hartwig, 25, Gunner's Mate 2nd Class
Michael William Helton, 31, Legalman 1st Class
Scott Alan Holt, 20, Seaman Apprentice
Reginald L. Johnson Jr., 20, Seaman Recruit
Brian Robert Jones, 19, Seaman
Nathaniel Clifford Jones Jr., 21, Seaman Apprentice
Michael Shannon Justice, 21, Seaman
Edward J. Kimble, 23, Seaman
Richard E. Lawrence, 29, Gunner's Mate 3rd Class
Richard John Lewis, 23, Fire Controlman, Seaman Apprentice
Jose Luis Martinez Jr., 21, Seaman Apprentice
Todd Christopher McMullen, 20, Boatswain's Mate 3rd Class
Todd Edward Miller, 25, Seaman Recruit
Robert Kenneth Morrison, 36, Legalman 1st Class
Otis Levance Moses, 23, Seaman
Darin Andrew Ogden, 24, Gunner's Mate 3rd Class
Ricky Ronald Peterson, 27, Seaman
Mathew Ray Price, 20, Gunner's Mate 3rd Class
Harold Earl Romine Jr., 19, Seaman Recruit
Geoffrey Scott Schelin, 20, Gunner's Mate 3rd Class
Heath Eugene Stillwagon, 21, Gunner's Mate 3rd Class
Todd Thomas Tatham, 19, Seaman Recruit
Jack Ernest Thompson, 22, Gunner's Mate 3rd Class
Stephen J. Welden, 24, Gunner's Mate 2nd Class
James Darrell White, 22, Gunner's Mate 3rd Class
Rodney Maurice White, 19, Seaman Recruit
Michael Robert Williams, 21, Boatswain's Mate 2nd Class
John Rodney Young, 21, Seaman
Reginald Owen Ziegler, 39, Senior Chief Gunner's Mate

President George H. W. Bush and Capt. Fred Moosally after the funeral service of April 24.
NARA

Caskets containing the bodies of the victims arrive at Dover Air Force Base.
NARA

Capt. Larry Seaquist, formerly in command of *Iowa*, explains the unfolding of the catastrophe to the press.
USN

Vice Adm. Richard D. Milligan during a press conference in the Pentagon.
USN

The loading of a 16-inch gun.
USN

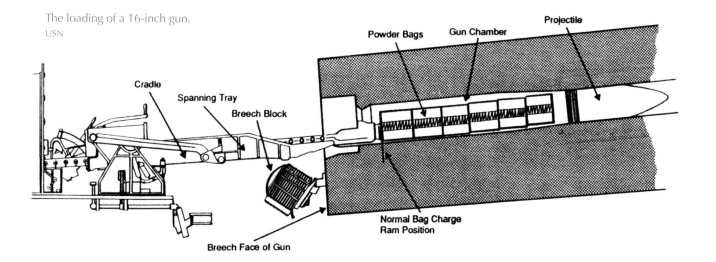

Iowa sails for Norfolk Naval Shipyard.
NARA

Despite the trauma suffered by her officers and men, *Iowa* resumed her operational career and proceeded to Whisky anchorage off Norfolk for rearming between May 30 and June 2, 1989.

Iowa's last cruise began on June 7 and took her to Northern Europe and then the Mediterranean, with stops at Kiel, Portsmouth, Rota, Casablanca, Gibraltar, Marseille, Antalya, Gaeta, Istanbul, Haifa, Alexandria, Ajaccio, Augusta Bay, Naples, and Palma. From September 16 she participated in Exercise Display Determination before parting company with the Sixth Fleet on November 26. She arrived at Norfolk on December 7 having fired her last heavy-caliber shells during the voyage.

In May Captain Moorse briefly became the twentieth and last commander of the vessel, which offloaded her last ammunition on October 26. She had fired 11,834 16-inch shells since entering service in February 1943.

Iowa was stricken from the Navy List in January 1995, but the following year Section 1011 of the National Defense Authorization Act required the Navy to keep two units of the class in readiness in the event of force majeure.

In September 1998 *Iowa* was towed to the Naval Education and Training Center at Newport, Rhode Island, where she served as a training ship before again traversing the Panama Canal in March 2001 on her way to joining the Reserve Fleet at Suisun Bay near San Francisco. A decade at anchor awaited her.

A souvenir photo of part of the crew mustered on the forecastle.
USN

The forecastle with its formidable 16-inch armament.
USN

Iowa leading a powerful battle group consisting of, among other vessels, the carrier *Saratoga* (CV 60) to port and *Coral Sea* (CV 43) to starboard.
NARA

Iowa and Old Glory.
USN

Iowa sailing from the port of Marseille.
PRADIGNAC AND LEO

Shadowing a Soviet Krivak I-class guided-missile frigate.
NARA

Iowa passing under Newport Bridge on September 24, 1998, on her way to the Naval Education and Training Center at Newport, Rhode Island.
USN

Iowa as part of the Reserve Fleet at Suisun Bay near San Francisco.
USN

Iowa's last voyage.
USN

Iowa tied up at San Pedro.
PHILIPPE CARESSE

Floating Museum in Los Angeles

On January 4, 1999, Congress announced that *Iowa* alone would remain on armed standby as the Navy wanted to keep a single unit operational to provide fire support in the event of a troop landing on a hostile shore. The federal government therefore required the entire class to be kept in reserve.

After many years of immobilization it was eventually decided to transform *Iowa* into a naval memorial, the last of her sisters to be so converted. The deterioration of the ship, however, was such that the restoration was estimated at between $15 and $20 million, meaning that a budget of $4 million and 125,000 visitors per year would be required to make the project viable. It remained to find the ideal spot to put her on display.

In March 2007, the Historic Ships Memorial at Pacific Square, Vallejo, site of the former Mare Island Naval Shipyard, and a Stockton-based group submitted proposals to convert the ship into a floating museum. These were approved by Iowa Senate Resolution No. 19 on April 25, 2009, but in February 2010 the Pacific Battleship Center made an offer to berth *Iowa* at San Pedro in the city of Los Angeles. This was endorsed on April 12 by Iowa governor Chet Culver, who pledged his support to complete the project to the tune of $10 million. On May 24 the Federal Register announced that the vessel would be transferred to a city or not-for-profit organization in California. On November 18, the Port of Los Angeles declared that Berth 87 could be allocated for mooring the ship and her development as a naval memorial. On September 6, 2011, management of *Iowa* was awarded to the Pacific Battleship Center, which as a not-for-profit organization would not receive any funding from the City or County of Los Angeles. The State of Iowa, however, contributed to the preservation of the vessel with a $3 million donation for preparation of the vessel. The Pacific Battleship Center therefore relies on admissions, memberships, and donations to keep *Iowa* open to the public.

In October 2011 the vessel was towed to Richmond, California, for refitting and a complete coat of paint, being opened to the public on weekends.

On April 30, 2012, Vice Adm. W. Mark Skinner signed the vessel over to Robert Kent, president of the Pacific Battleship Center, although the U.S. Navy continues to regard *Iowa* as being in reserve. After a long tow to the Port of Los Angeles that began on May 26 and a period at anchor off the Southern California coast for removal of any invasive species or contaminants from her hull, *Iowa* was brought in to Berth 87 at San Pedro on June 9 and opened to the public on July 7.

Approaching San Pedro, the superstructure of *Iowa* can be seen rising above the port where the ship is moored starboard to the dock. Boarding the ship by the forward gangway, the first sight is of her bell under the dicone/discage antenna and the forecastle dominated by the bulk of the two 16-inch turrets. Entering the ship one can visit the senior officers' cabins, washrooms, and the officers' wardroom as well as the suite assigned to President Roosevelt that still houses the bathtub specially installed for his use. Further discoveries await in the form of the enlisted men's galley, laundry, barbershop, and crew berthing. Touring the main deck, visitors pass by the 5-inch mounts, and a series of ladders leads to the bridge and conning tower, still fully equipped. Access to the engine spaces, magazines, and battery plot compartments is available by special arrangement.

David Way, the genial and generous curator of the Battleship *Iowa* Museum.
PHILIPPE CARESSE

USS *IOWA* (BB 61)

Commissioned	February 22, 1943
Decommissioned	March 24, 1949
Commissioned	August 25, 1951
Decommissioned	February 24, 1958
Commissioned	April 28, 1984
Decommissioned	October 26, 1990

Pacific Battleship Center
250 S. Harbor Blvd.
Berth 87
Los Angeles, CA 90731
USA
www.pacificbattleship.com

The superstructure of *Iowa* with her two forward 16-inch turrets.
RAINER SCHITTENHELM

Veterans provide visitors with a living connection to the career of the ship.
PHILIPPE CARESSE

A corner of one of the engine rooms.
PHILIPPE CARESSE

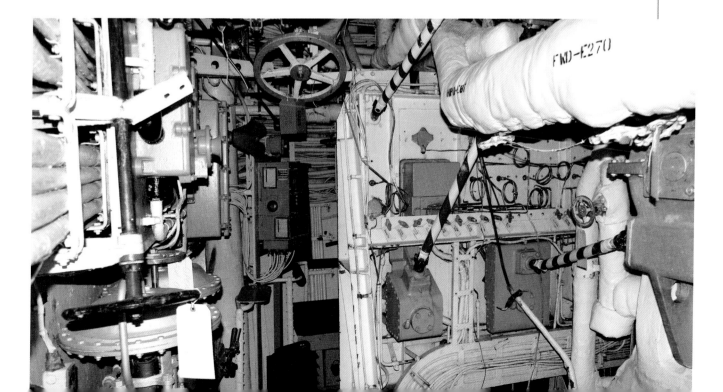

The officers' wardroom.
PHILIPPE CARESSE

The central passageway called "Broadway" leading to the engine rooms and the firerooms.
PHILIPPE CARESSE

Iowa is in a remarkable state of preservation.
RAINER SCHITTENHELM

A view alongside the bridge toward the forecastle.
PHILIPPE CARESSE

The superstructure and turret 3 seen from the fantail.
PHILIPPE CARESSE

FOURTEEN

The Career of the Battleship USS *New Jersey* (BB 62)

Reconquering the Pacific

The keel of *New Jersey* (BB 62) was laid at Philadelphia Naval Shipyard on September 16, 1940, and the ship was launched by Mrs. Carolyn Edison, wife of the governor of New Jersey, on December 7, 1942. The event was attended by more than 20,000 people, and the hull entered the water with a displacement of 36,447 tons a little after 1415 that afternoon.

Completion of the work was supervised by naval construction engineers Francis Forest and Allan Dunning at the head of eight thousand workers of all types. The 16-inch guns were among the first of her weapons to be installed together with the forward superstructure and rangefinders. A fire broke out during the installation of the conning tower but was quickly brought under control.

New Jersey was commissioned on May 23, 1943, under the command of Capt. Carl Holden after a construction period of forty-one months. The vessel remained at Philadelphia until July as crewmen and technicians spared no efforts to prepare her for sea at the earliest opportunity.

Trials began on July 8 with a first run in Delaware Bay. These were a success and the first Kingfisher floatplane was launched on the 20th. Two days later Undersecretary of the Navy James Forrestal came on board to observe the trials and first gunnery exercises, during which she worked up to 32.4 knots.

The end of the month found *New Jersey* moored at Annapolis, the crew having already christened her "The Big J," the name that attached to her for the rest of her long career. Another early arrival was the ship's canine mascot Spike.

The hull of *New Jersey* taking shape in the summer of 1942.
NARA

New Jersey under construction in the Philadelphia Naval Shipyard on July 8, 1942.
USN

New Jersey a hive of activity in the summer of 1942.
NARA

Launch day for *New Jersey*, December 7, 1942.
NARA

The shafts await their propellers and the rudder bearings their rudders.
NARA

New Jersey launched at the Philadelphia Naval Shipyard on December 7, 1942.
USN

New Jersey shortly after launching.
USN

Turret 2 being swung into position on January 12, 1943.
USN

The commissioning service for *New Jersey*, May 23, 1943.
USN

New Jersey puts to sea to carry out her first trials, August 5, 1943.
USN

New Jersey with her newly enclosed bridge and its large bay windows.
USN

New Jersey leaves Philadelphia Naval Shipyard on August 7, 1943.
USN

On August 9 *New Jersey* sailed for her steam trials in the Gulf of Paria and off the Atlantic coast of Venezuela escorted by the destroyers *Sproston* (DD 577) and *Charrette* (DD 581). An intense shakedown cruise awaited her crew, with refueling at sea from the oiler *Mattole* (AO 17) and simulated torpedo attacks by aircraft from the light carrier *Monterey* (CVL 26).

On the 24th synchronization trials were carried out between the main battery and the plotting rooms. Unfortunately, G2 Irving S. Roremus found himself in the well of the center gun of turret 1 when it elevated suddenly. Unable to escape in time, he was crushed by the breech and died a few hours later.

On the 31st the battleship turned for Norfolk, where she was inspected by Rear Adm. Donald B. Beary on September 5 and 6. The following day she sailed for Philadelphia to be docked. She was refloated on October 3 and embarked 97,920 20-mm shells and 107,488 40-mm shells. On the 12th she was in Chesapeake Bay and was at Hampton Roads on the 17th to rectify some technical issues. The following day Captain Holden was able to inform the commander in chief of the Atlantic Fleet, Adm. Royal E. Ingersoll, that his ship was ready for battle, and that same day the crew was given five days' leave from Norfolk.

Launch trials for one of the Kingfisher floatplanes.
USN

New Jersey during her steam trials in the Gulf of Paria.
USN

New Jersey shortly before being assigned to the Pacific.
NARA

The fantail on September 3, 1943, with its three Vought OS2U Kingfisher floatplanes.
USNHC

New Jersey anchored at Hampton Roads on September 7, 1943, with the French battleship *Richelieu* beyond.
USN

New Jersey seen from *Iowa*.
USN

New Jersey's first mission was to sail on October 24th to Casco Bay, Maine, escorted by five destroyers as Task Force 22 to relieve her sister ship *Iowa* at Argentia, Newfoundland. Her presence in these waters was to counter any attempt by the German battleships *Tirpitz* and *Scharnhorst* to break out into the Atlantic. But with *Tirpitz* seriously damaged by British midget submarines on September 22, and the Royal Navy capable of handling a sortie by the *Scharnhorst*, *New Jersey* turned south and lay at Boston between December 16 and 18. At the end of that year she embarked 324 16-inch shells at Hampton Roads before carrying out maneuvers with *Iowa* from January 2, 1944. These two battleships then traversed the Panama Canal with a stop at Balboa on the 7th and 8th. They were intended for the U.S. Fifth Fleet, being attached to the 7th Battleship Division of Task Group 58.3 under the command of Rear Adm. Olaf M. Hustvedt, which was in turn part of Task Force 58 led by Adm. Marc A. Mitscher.

The two vessels reached Funafuti Atoll in the Gilbert and Ellice Islands on January 22, where they joined the force commanded by Rear Adm. Forrest Sherman flying his flag in the carrier *Bunker Hill* (CV 17). Sherman bluntly informed the commanders of both vessels that priority was given to antiaircraft defense and that he would gladly dispense with the services of their heavy guns. Within twenty-four hours of their arrival, *New Jersey* and *Iowa* were sailing north with *Bunker Hill*, the escort carriers *Cowpens* (CVL 25) and *Monterey* (CVL 26), the heavy cruiser *Wichita* (CA 45), and nine destroyers. At 0300 on the 29th more than seven hundred U.S. aircraft took off to attack the Marshall Islands in connection with Operations Flintlock and Catchpole aimed at capturing Kwajalein, Majuro Atoll, and Eniwetok. *New Jersey* remained on standby with her group between Bikini and Eniwetok, but the Japanese fleet never appeared and on February 4 Task Group 58.3 dropped anchor off Majuro.

New Jersey with Iowa beyond photographed by an aircraft from the carrier Bunker Hill (CV 17).
NARA

Five days later the heavy cruiser *Indianapolis* (CA 35) came alongside and transferred Vice Adm. Raymond Spruance on board *New Jersey*, in which he hoisted his flag as commander in chief of the Fifth Fleet. Joining *New Jersey* and *Iowa* as part of Task Group 50.9 were the heavy cruisers *Minneapolis* (CA 36) and *New Orleans* (CA 32) and the destroyers *Burns* (DD 588), *Izard* (DD 589), *Charrette* (DD 581), and *Bradford* (DD 545).

On February 17, during the course of the assault on Truk (Operation Hailstone), the *Iowa* class had its one and only experience of combat with surface units. In mid-morning, *New Jersey*'s Kingfisher spotted enemy vessels twenty-five miles to the northwest. Conditions were perfect with excellent visibility and light winds. After increasing speed, the Japanese squadron, consisting of the light cruiser *Katori*, the destroyers *Maikaze* and *Nowaki*, and the armed trawler *Shonan Maru No. 15*, was sighted at a range of 32,900 yards. Battle was joined at midday and *Katori*, already damaged by aerial attack, was quickly sunk by *Iowa*, though not before firing a salvo of torpedoes that passed between the two battleships. *New Jersey* engaged *Maikaze* with her 5-inch guns at a range of 7,650 yards, to which *New Orleans*, *Minneapolis*, *Burns*, and *Bradford* added their fire. *Maikaze* fought gamely until she blew up and sank with all hands at 1343 in position 07°45′ N, 151°20′ E. *Shonan Maru* did not hold out for long under *New Jersey*'s 5-inch and 40-mm fire, disappearing in a cloud of smoke. Only *Nowaki* was able to escape her pursuers, chased by salvos of 16-inch shells fired at a range of 38,000 yards. Admiral Spruance ordered fire to be checked after the range had exceeded 38,250 yards.

New Jersey and *Iowa* had acquitted themselves well during their first battle, but the main command position was found to have insufficient space, and radar suffered interference from VHF communications.

New Jersey's heavy guns scored telling hits against Japanese units during the battle off Truk on February 17, 1944.
NARA

The crew of a 40-mm Bofors mount take the opportunity to rest at their combat station.
USN

After rejoining Task Group 58.3, the ships dropped anchor off Kwajalein on the 19th. Two days later Spruance was promoted to admiral and duly bought his stars from the ship's store. Meanwhile, *New Jersey* was briefly attached to Task Group 58.2 under Rear Adm. Alfred E. Montgomery, and the 25th found her off Majuro. A spell of exercises followed involving *New Jersey*, *Iowa*, the carrier *Lexington* (CV 16), and four destroyers.

On March 17, the task group sailed for Mili Atoll, with Admiral Spruance having temporarily transferred to Pearl Harbor and command being assumed by Rear Adm. Willis A. Lee flying his flag in *Iowa*. The following day the battleships opened fire on fortified positions at a range of 15,000 yards, and the Japanese responded with 120-mm guns that inflicted two hits on *Iowa*. The damage was superficial and the shelling resumed once the battleships had shifted the range to 19,700 yards.

By the 19th the fleet was back at Majuro and Spruance reunited with his flagship, but the battleships of Task Force 58 were soon in action supporting the air raids on the Caroline Islands as part of Operation Desecrate I. The U.S. Navy mustered a significant force for this offensive consisting of six battleships, eleven heavy cruisers, three light cruisers, and twenty-four destroyers supporting the eleven aircraft carriers assigned to deliver air raids on Palau. This new target lay two thousand miles west of the point of departure, and Task Force 58 came under Japanese air attack during the approach on the night of the 29th. *New Jersey* was the first to detect the impending air attack with radar at a range of 30,000 yards. The SK radar equipment then took over and the 5-inch guns opened fire at around 2100 with an aircraft apparently shot down. The 40-mm and 20-mm soon joined in though unassisted by radar on this occasion. The attack ended as soon as it had begun and the fleet proceeded unchallenged. After the success of the Palau operation it was the turn of Woleai and its modest airfield to suffer the wrath of U.S. naval ordnance. The fleet returned to Majuro on April 6, with Admiral Spruance striking his flag in *New Jersey* before Task Force 58 moved on to its next objective.

On April 13, the fleet raised anchor and turned for New Guinea to participate in Operation Persecution, with *New Jersey* serving as part of Task Group 58.1 under Rear Adm. Joseph J. Clark. After refueling north of the Admiralty Islands, the force reached Hollandia on the 21st with the intention of covering the landings planned for the following day.

The magnificent silhouette of *New Jersey*.
NARA

New Jersey at sea.
USN

Repeated shelling of land targets required frequent replenishment of heavy-caliber shells.
USN

With naval operations proceeding uneventfully, Admiral Mitscher decided to carry out a new raid on Truk to empty the bomb bays of his embarked aviation. Humboldt Bay and Tanahmerah Bay were also subjected to bombardment. On May 1 *New Jersey* joined Task Group 58.2 and participated with *Iowa*, *North Carolina* (BB 55), *South Dakota* (BB 57), *Indiana* (BB 58), *Massachusetts* (BB 59), and *Alabama* (BB 60) in the destruction of the airfield at Ponape together with numerous shore installations in the vicinity. "The Big J" fired 90 16-inch and 154 5-inch shells. One of *New Jersey*'s gunners was taken ill with appendicitis during the attack, leaving the ship's surgeon no option but to intervene immediately, pausing each time a salvo was fired.

The battleship *South Dakota* anchored at Ulithi in the Caroline Islands in June 1944 with *New Jersey* beyond.
NARA

New Jersey arriving at Pearl Harbor on August 9, 1944.
NARA

The fleet returned to Majuro on May 4 and ten days later Vice Admiral Lee hoisted his flag in *New Jersey*. He did so only reluctantly as his favorite ship was the battleship *Washington*, which had unfortunately been in collision with the *Indiana* and was under repair. Denied the use of *Washington*'s sister *North Carolina*, which was obliged to sail for Pearl Harbor for rudder faults to be addressed, Lee had no option but to shift his flag to *New Jersey* until the 30th of that month.

On June 6 *New Jersey* and her group sailed for the Marianas as part of Operation Forager, the conquest of Saipan and Tinian. During the night of the 12th and the 13th the fleet was the target of a concerted air attack while steaming at 20 knots. General quarters was sounded at 0032 but the crew stood down at 0145 with the radar screens empty. An enemy aircraft was detected at 0333, however, and the Task Group again came to general quarters. At around 0400 a Mitsubishi G4M Betty torpedo bomber reached the fleet undetected at an altitude of five hundred feet and launched its torpedo eight hundred yards from *New Jersey*. This was, however, avoided by the battleship, and the Betty was promptly shot down by the Carrier Air Patrol (CAP).

The Fifth Fleet was responsible for protecting the 644-strong invasion fleet, and it did not go untested since the Japanese attacked with no fewer than 350 aircraft on the 19th. *New Jersey* was hard put to repel waves of Mitsubishi A6ME Zero, Nakajima B6N Tenzan Jill, and Kawasaki Ki-61 Hien Tony aircraft, but the Japanese suffered heavy losses at the hands of F6F Hellcats. Bombardments were carried out at a range of 9,850 yards against Japanese fortified positions and ammunition dumps. After landing force was put ashore, the fleet withdrew to Guam to refuel before setting out in pursuit of Vice Admiral Ozawa Jisaburō's fleet of four carriers, two hybrid battleships, three heavy cruisers, and eight destroyers. No contact was made, however, and *New Jersey* rejoined Rear Adm. John W. Reeves' Task Group 58.3 on the 24th. Early the following afternoon one of *New Jersey*'s Kingfisher floatplanes flew to Guam to collect three aviators from the *Lexington* (CV 16). The island still being in Japanese hands, the Kingfisher received several hits and struggled to take off thanks to her additional payload and was forced to put down near the destroyer *Caperton* (DD 650) to transfer its passengers. It then returned to *New Jersey* with no further misadventures.

Except for a brief stop at Saipan, *New Jersey* spent the whole of July escorting the aircraft carriers. Now part of Task Group 58.4, she steamed off Guam, which was shelled on several occasions. On August 4 she was off Eniwetok making preparations to return to Pearl Harbor to receive Adm. William F. Halsey, commander in chief of the Third Fleet. She reached Hawaii on the 9th, where the crew was given leave while the ship's electrical circuitry was refitted. The ship replenished and refueled, and the arrival of Admiral Halsey on the 24th formed the prelude to a series of night exercises with destroyers. On the 28th an exercise flight by one of the ship's Kingfishers resulted in the plane ditching two thousand yards to starboard of the ship. The pilot and observer were rescued by the destroyer *Hickox* (DD 673), but Halsey, unwilling to waste time recovering the still-floating aircraft, ordered it sent to the bottom.

Another view of *New Jersey* as she approaches Pearl Harbor on August 9, 1944.
NARA

On August 30 *New Jersey* sailed from Pearl Harbor for Manus in the Admiralty Islands, where she anchored on September 4. The following day she sailed with Task Group 38.5 to join TG 38.2 and with it *Iowa*. The rendezvous on the 9th was followed by refueling from the oilers *Lackawanna* (AO 40) and *Cimarron* (AO 22). Aside from *New Jersey* and *Iowa*, Task Group 38.2 now consisted of the aircraft carriers *Intrepid* (CV 11), *Bunker Hill* (CV 17), and *Hancock* (CV 19), the escort carriers *Independence* (CVL 22) and *Cabot* (CVL 28), the light cruisers *Vincennes* (CL 64), *Miami* (CL 89), and *Oakland* (CL 95), and eighteen destroyers.

On the 12th, the destroyer *Marshall* (DD 676) came alongside to transfer survivors of the light cruiser *Natori*, sunk after being torpedoed in the San Bernardino Strait by the submarine *Hardhead* (SS 365) on August 30. Escort duty continued during the raids against Luzon as well as the bombardment of Shinihi and Macujama, before the ship put in at Garapan on Saipan.

By early October Task Force 38 was preparing for Operation King II, the invasion and seizure of Leyte in the Philippines. Not only was control of this archipelago vital to Japan's strategic position in Southeast Asia, but it was also a source of raw materials essential to the Japanese war effort. On October 20, more than 120,000 U.S. troops landed on Leyte. That same day the Japanese launched Operation Shō-Ichi-Gō Sakusen, which aimed to cut the invasion fleet off from Admiral Halsey's main force before annihilating the barges and other amphibious landing vessels. The main instrument of the plan was the fleet led by Vice Admiral Kurita Takeo, which consisted of three battleships, six heavy cruisers, one light cruiser, and nine destroyers. This battle group was required to force the San Bernardino Strait by night while Vice Admiral Ozawa attempted to draw the U.S. aircraft carriers away to the north. Things did not, however, go to plan and Kurita lost three of his heavy cruisers to U.S. submarines on the 23rd with another forced to return to Brunei. Lacking any form of air cover, the following day the giant battleship *Musashi* was overwhelmed by U.S. carrier aircraft and sank that evening. Dismayed by this turn of events, Kurita eventually ordered his squadron to reverse course to regain the open water.

Informed of these developments, Admiral Halsey was anxious to deliver the coup de grâce by pursuing the fleeing Japanese through the San Bernardino Strait. But *New Jersey*, *Iowa*, the light cruisers *Vincennes*, *Miami*, and *Oakland*, and eight destroyers failed to catch the Japanese fleet and instead came under air attack. In the early hours of the 25th the U.S. forces turned south and obtained a radar surface contact. No vessel was found despite intensive efforts, but Japanese records subsequently revealed this to have been the destroyer *Nowaki*, which had escaped the attentions of *New Jersey* on February 17 and was eventually sunk by the destroyer *Owen* (DD 536) on the 26th. On the 29th *New Jersey* beat off a kamikaze attack and destroyed an aircraft preparing to strike the carrier *Intrepid*.

On November 4, *New Jersey* was in collision with the destroyer *Colahan* (DD 658) during a mail transfer operation, but damage to both ships was slight and there were no casualties. On the 9th, Task Group 38.2 put in at Ulithi in the Caroline Islands for a well-deserved rest, but the ships were at sea again five days later to relieve TG 38.3 off Luzon. It was at this time that the kamikaze attacks increased in intensity, testing the morale and resolve of the U.S. sailor to the limit. On the 25th three Zekes were shot down by *New Jersey* although their targets were the carriers *Intrepid*, *Hancock*, and *Cabot*. Despite the effectiveness of their antiaircraft armament, numerous ships were hit and severely damaged.

The giant Japanese battleship *Musashi* sinking by the head.
USN

Admiral Kurita's fleet sails from Brunei Bay on October 22, 1944, as part of the operation that led to the Battle of Leyte Gulf.
USN

Task Force 38 anchored at Ulithi. A hospital ship is in the foreground, possibly *Samaritan* (AH 10), with two battleships of the *Iowa* class and another of the *South Dakota* class seen beyond with an assortment of light and fleet carriers in the background.
USN

The view from *New Jersey* as the carrier *Intrepid* (CV 11) is struck by a kamikaze on November 25, 1944.
USN

New Jersey under attack by kamikaze aircraft in a photo taken from the flight deck of the carrier.
USN

New Jersey and the carrier *Hancock* (CV 19) weather a typhoon in November 1944.
USN

Despite the gravity of these events, morale remained good in *New Jersey*. Admiral Halsey often visited the men at their quarters, would not hesitate to join the crew of a turret for a cup of coffee, and shared Thanksgiving dinner with the men on November 23, 1944. He disembarked on January 27, 1945, to be replaced two days later by Rear Adm. Oscar C. Badger.

Meanwhile, *New Jersey*'s sister *Wisconsin* had reached Ulithi on December 9 to reinforce the Third Fleet. On the 11th, Task Group 38.2 sailed for Luzon but a week later were assailed by Typhoon Cobra (known as "Halsey's Typhoon") while still three hundred miles short of its destination. The destroyers *Hull* (DD 350), *Monaghan* (DD 354), and *Spence* (DD 512) foundered with heavy loss of life in a sixty-five-foot swell and winds in excess of 100 knots, which eventually claimed 790 lives. *New Jersey*, rolling through 20 to 25 degrees, had one of her Kingfishers washed away. Once calm had been restored, TG 38.2 refueled 250 miles east of the Philippines before returning to Ulithi on the 24th. The holiday season was spent at anchor, and on January 26, 1945, Capt. Edmund Wooldridge took command of *New Jersey*, with the following days being spent in antiaircraft exercises.

Admiral Halsey joins his sailors for a meal in *New Jersey*. USN

Task Group 58.3 sails from Ulithi in 1945.
USN

On February 10, *New Jersey* was assigned to Task Group 58.3, where she joined the battleship *South Dakota*, the heavy cruiser *Indianapolis*, and the light cruisers *Pasadena* (CL 65), *Astoria* (CL 90), and *Wilkes-Barre* (CL 103). Together with eighteen destroyers, these vessels were responsible for escorting the carriers *Essex* (Vice Adm. Forrest P. Sherman), *Bunker Hill*, and *Cowpens* (CVL 25). Operation Jamboree began that same day with the ships preparing to cover the intended landings on Iwo Jima. On the day of the Marine assault, February 19, *New Jersey* was lying northwest of Mount Suribachi awaiting any Japanese counterattack. On the 25th she covered the carrier raids on Tokyo and returned to Ulithi on March 5 for nine days of rest. On the 14th, Task Group 58.3 sailed for Kyūshū to protect Task Force 58 on its bombardment missions on Japan's southernmost island. At 0715 on the 19th, however, squadrons of Japanese aircraft appeared over the fleet resolved to attack the aircraft carriers. *Franklin* (CV 13) was hit by two bombs while at 0741 *New Jersey* shot down a Yokosuka D4Y Judy dive bomber attempting to strike the *Essex*. A second Judy was destroyed attempting to reach the *Bunker Hill* at 0813, and determined attacks were repulsed throughout the day.

On the 22nd *New Jersey* was assigned to Rear Adm. Arthur W. Radford's Task Group 58.4 and refueled 600 miles off Kyūshū. In the early hours of the 24th she joined *Iowa* and *Wisconsin* in pounding Kutaka Shima off Okinawa and that afternoon hit the southeast coast of Okinawa itself. April 1 saw the start of Operation Iceberg, consisting of a major landing on the shores of Haguchi Bay on Okinawa. The following day TG 58.4 turned south to refuel. At approximately 2112 that night, as the ships proceeded in company with the flagship *Yorktown* (CV 10), the destroyer *Franks* (DD 554) set a converging course and collided with *New Jersey* at 23 knots. Damage to *New Jersey* was slight, but the same was not true of *Franks*, which suffered the destruction of her superstructure and part of her port armament and lost her commanding officer, Cdr. David R. Stefan. The vessel had to make her way to Puget Sound for repairs.

Between March 24 and April 19 a bombardment force was established as Task Group 58.7 consisting of *New Jersey*, *Missouri*, *Wisconsin*, *South Dakota*, *North Carolina*, *Massachusetts*, *Indiana*, and *Washington*.

Its role was mainly that of shelling shore installations in areas planned for troop landings, but on April 6 a B-29 reported that the battleship *Yamato* (Vice Admiral Itō Seiichi) and an escort consisting of the light cruiser *Yahagi* and eight destroyers was under way in the Bungo Strait. This movement, called Ten-Ichi-Gō, heralded the final sortie of the Japanese fleet, a suicide mission in which the last of its giant battleships would attempt to put herself aground on the coast of Okinawa in an effort to repel the U.S. invasion. There was no escape for these vessels, which had fuel only for the outward voyage.

On the morning of the 7th, six battleships of TG 58.7, including the three *Iowa*s, were standing by fifty miles off Okinawa to intercept *Yamato* if she survived destruction by aircraft from Task Force 58. From 1027 that morning the Japanese squadron suffered repeated attacks from Hellcat, Helldiver, Avenger, and Corsair aircraft from the carriers *Hornet* (CV 12), *Bennington*, *Bunker Hill* (CV 17), *San Jacinto* (CVL 30), *Essex* (CV 9), *Yorktown*, *Intrepid* (CV 11), *Langley* (CVL 27), and *Bataan* (CVL 29). Struck by an estimated eight heavy bombs and fourteen torpedoes, *Yamato* succumbed with much of her crew.

The destroyer USS *Franks* (DD 554).
USN

Damage to *Franks* after her collision with *New Jersey*.
USN

New Jersey's armament would never be used for the purpose for which it was designed, namely an encounter with a battleship, of which *Yamato* would have been a worthy opponent. Rear Adm. William Abhau, then a gunnery officer, aptly summarized her wartime record: "We were part of the antiaircraft protection. We did a good deal of shore bombardment. We also served as an underway refueling ship and a hospital ship. In fact, we did everything except be a battleship."

There was still time, however, for *New Jersey* to shoot down a Zeke on the 11th, and three days later the crew was informed that the ship was returning home. After reaching Ulithi, where she was relieved by *Iowa* on the 16th, she set a course for Pearl Harbor accompanied by the heavy cruiser *Minneapolis* (CA 36). She spent just twenty-four hours in Hawaii before heading for Bremerton, reaching Puget Sound Naval Shipyard on May 6. After landing her ammunition she was docked for refitting.

The great Japanese battleship *Yamato*.
KURE MARITIME MUSEUM

New Jersey refueling at sea on April 9, 1949. Beyond her is *Wisconsin* and an *Essex*-class carrier.
USN

New Jersey tied up at Puget Sound Naval Shipyard.
USN

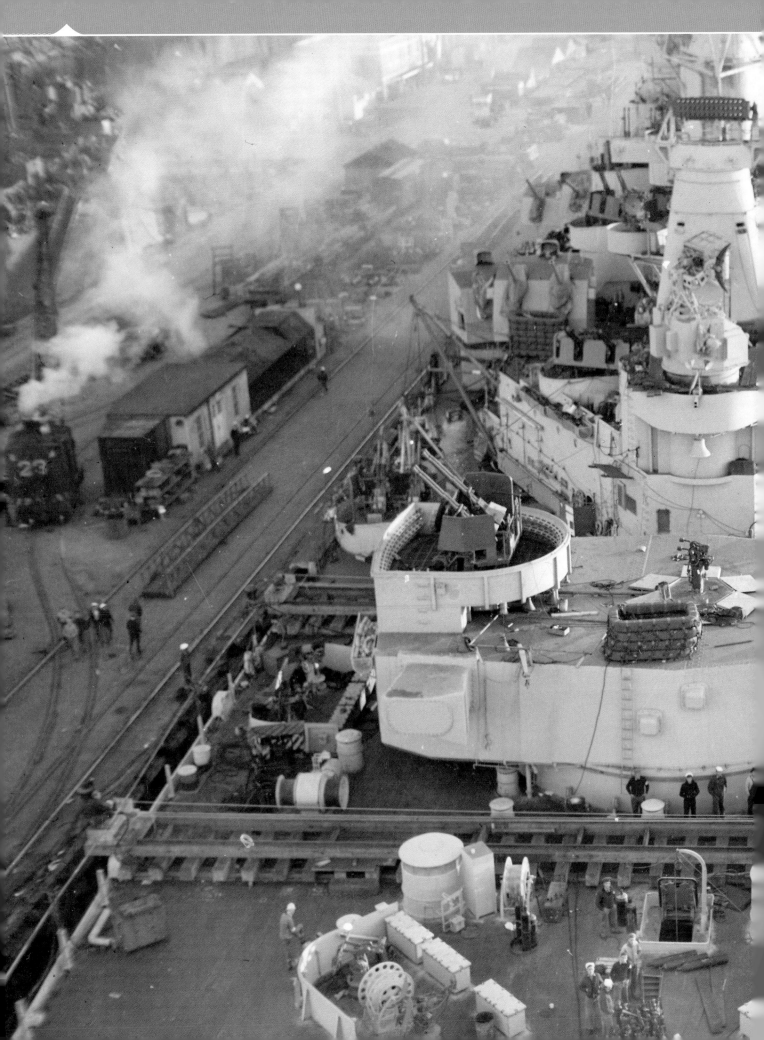

New Jersey tied up at Puget Sound Naval Shipyard.
USN

New Jersey tied up at Puget Sound Naval Shipyard.
USN

On June 30, *New Jersey* resumed training and on July 4 sailed for San Pedro with the light cruiser *Biloxi* (CL 80) and the destroyer *Norris* (DD 859). On the 19th she turned for Pearl Harbor where she made a brief stop before heading for Wake Island on August 8 for a bombardment mission. The following day she reached Eniwetok with *Biloxi* and four destroyers, and on the 10th the ammunition ship *Durham Victory* came alongside to complete a transfer. Meanwhile, momentous events were in train, and on the 14th the Japanese Imperial Council requested unconditional surrender. On the 20th Admiral Halsey shifted his flag from *Missouri* to *New Jersey*, which entered Manila in heavy rain on the 21st to find the city 85 percent destroyed. A week later *New Jersey* anchored in Buckner Bay, Okinawa, weighing on September 13 for the port of Wakayama on the 15th before entering Tokyo Bay on the 17th.

In November Capt. Edward Thompson was appointed in command, and on the 28th *New Jersey* was at Yokosuka with Adm. John H. Towers embarked as flagship of the Fifth Fleet, with the ships anchored near the damaged Japanese battleship *Nagato*. *New Jersey* quit her anchorage only briefly to cruise off the coast with the light cruiser *Pasadena* (CL 65).

New Jersey anchored at Puget Sound on July 2, 1945.
USN

New Jersey in Tokyo Bay on December 30, 1945. The Japanese battleship *Nagato* can be seen in the background.
USN

Towers left the ship on January 18, 1946, and *New Jersey* headed for the United States on the 29th after being relieved by *Iowa*. Embarked in her were thousands of troops for repatriation as part of Operation Magic Carpet, who disembarked in San Francisco on February 10. On March 26 *New Jersey* proceeded to the Long Beach Naval Shipyard, where a fresh crew came on board. Here she remained throughout the spring carrying out various sorties and exercises with *Iowa* under Rear Adm. John W. Roper before both vessels sailed for Puget Sound on May 29. On June 1 some 1,000 16-inch shells, 11,000 5-inch shells, 100,000 40-mm shells, and 85,000 20-mm shells were embarked at Bangor, Washington. *New Jersey* spent the next few months inactive, the highlight coming on October 24, when the freighter *Sylvania* was driven into the hull of the battleship by a powerful current. Neither ship suffered other than slight damage, which in the case of *New Jersey* extended to the breakage of a few light bulbs.

On January 11, 1947, *New Jersey* became subject to Inspection and Survey (InSurv) and entered a floating dock with a reduced crew of eight hundred men for the hull to be examined. Under ordinary circumstances the ship would have passed into the reserve, but a new mission awaited her, that of a training vessel for midshipmen. *New Jersey* emerged from docking on March 12 and on April 2 returned to Bangor to rearm. Three days later she sailed for the Caribbean via Long Beach and the Panama Canal, with gunnery exercises being carried out in Guantánamo Bay in late April and early May before the ship docked at Bayonne in her namesake state on the 13th. Ten days later she was visited by the governor of New Jersey to mark the fourth anniversary of her commissioning.

On the 24th, *New Jersey* headed for Norfolk to embark reservists and on June 3 was at Annapolis, where together with *Wisconsin* she received 518 midshipmen from the United States Naval Academy and the Naval Reserve Officers Training Corps (NROTC). With *Wisconsin* flying the flag of Rear Adm. Heber H. McLean, these two ships sailed for Cape Henry, Virginia, where they joined Task Force 81, composed of the carriers *Randolph* (CV 15) and *Kearsarge* (CV 33) and the destroyers *Cone* (DD 866), *Stribbling* (DD 867), *O'Hare* (DD 889), and *Meredith* (DD 890), for a training cruise in European waters. After an uneventful crossing of the Atlantic including refueling by the fleet oiler *Chemung* (AO 30), the squadron rounded Cape Wrath (Scotland) on the 22nd and entered the Firth of Forth the following day. Usually based in London, Adm. Richard L. Conolly, commander of U.S. Naval forces in the Atlantic and the Mediterranean, came on board at Rosyth, choosing to hoist his flag in *New Jersey* during the planned cruise of Northern Europe.

New Jersey on March 31, 1947.
USN

New Jersey towed past Manhattan on her way to join the Atlantic Fleet Reserve at Bayonne, New Jersey.
USN

On the 29th *New Jersey* and *Wisconsin* sailed for Oslo, where King Haakon VII came on board *New Jersey* while the rest of the squadron sailed for the Swedish port of Gothenburg. After reassembling in the North Sea, Task Force 81 headed for Portsmouth, England, where it arrived on July 9. Admiral Sir Bruce Fraser, victor of the Battle of North Cape and former commander in chief of the British Pacific Fleet, came on board and a score of cadets were invited to Buckingham Palace to attend a reception given by King George VI.

On the 18th, Admiral Conolly hauled down his flag and the fleet turned for Guantánamo Bay, where it arrived on August 1. After gunnery exercises targeting the island of Culebra off Puerto Rico, *New Jersey* landed her midshipmen at Annapolis on the 25th and made for Gravesend Bay, New York, to offload ammunition before going into reserve. On September 3 the lofty aerial of the SK-2 radar was removed so the vessel could clear the Brooklyn Bridge and make her way up the East River to the New York Naval Shipyard for deactivation. By January 1948 the work of decommissioning was in progress, with the ship's most sensitive equipment being put into mothballs. Towed back to Gravesend Bay for removal of her fuel, she was then brought to Bayonne and retired from active service on June 30, 1948, before being placed in the Atlantic Reserve Fleet.

At Sea in the Fifties

New Jersey was not destined to spend long in the reserve, as tensions in the Korean Peninsula resulted in U.S. intervention from June 1950. With *Missouri* still in commission, the U.S. Navy responded by reactivating the rest of the *Iowa* class to support troop movements ashore with their formidable armament.

Accordingly, workmen came on board *New Jersey* on September 26, 1950, followed by the crew in October. The ship's recommissioning ceremony took place on November 21, Capt. David Tyree being appointed in command. The following day sixteen tugs drew her to the floating dock at the New York Naval Shipyard. The task of returning her to service was not without incident, and several electrical fires had to be contained over the next two months, but on January 16, 1951, *New Jersey* was able to sail to Norfolk under her own power. Having taken on supplies of food and ammunition, she embarked on her steam trials in Guantánamo Bay followed by gunnery exercises from February 23. Returning to Norfolk at a speed of 32 knots on March 19, leave was given to the crew before she made for Portsmouth, Virginia, on April 6 prior to traversing the Panama Canal on the 20th. The following day she was joined at Balboa by her sister *Missouri* and the officers of both vessels took the opportunity to meet for a briefing on operations in the Korean theater.

Present at *New Jersey*'s recommissioning service on November 21, 1950, are her new captain, David Tyree, Governor Alfred Driscoll of New Jersey, Adm. William F. Halsey, and Vice Adm. Oscar C. Badger.
USN

Refitting at Bayonne shortly before being recommissioned into the Navy.
USN

New Jersey at sea in the 1950s.
USN

Mail being transferred between *New Jersey* and the destroyer *Trathan* (DD 530) during the Korean War.
USN

A detail of the photo below showing *New Jersey*'s main range-finder and detection equipment during the Korean War.
USN

After calling at Pearl Harbor on April 30, *New Jersey* sailed for Japan on May 4, reaching Yokosuka eight days later. On the 13th, Vice Adm. Harold M. Martin, commander in chief of the Seventh Fleet, hoisted his flag on board and two days later *New Jersey* sailed in company with the carrier *Philippine Sea* (CV 47), the light cruiser *Manchester* (CL 83), and three destroyers to join Task Force 77 off Korea. On the 19th, *New Jersey* and two destroyers were detached to shell Kansong and Wonsan. The first of these targets was taken under fire between 0500 and 0630 on the 20th, and the second from 2310 that day. At 0300, however, an incandescent particle ignited some 40-mm ammunition stowed in the forward superstructure. The blaze was quickly brought under control, but at 0932 the following day an enemy 4-inch or 5-inch shell struck turret 1, killing one man and wounding two more, an incident that gave Seaman Apprentice Robert Osterwind the unfortunate distinction of being the only battle fatality ever suffered in *New Jersey*. Within seven minutes of being struck, *New Jersey* had identified the position of the enemy battery on the Kalmagak Peninsula and quickly silenced it. Damage was also caused to the ship by the blast of her own gunfire, particularly that from turret 3.

Before the month was out, *New Jersey* had shelled Yangyang and Kansong for a second time, destroying a bridge and three ammunition dumps. On May 24 one of her helicopters was lost while en route to rescue a pilot from the carrier *Boxer* (CV 21). It was believed on board *New Jersey* that the crew had likely perished, but they managed to come ashore on friendly territory and make contact with the destroyer *Arnold J. Isbell* (DD 869), which returned them to the ship three days later.

After rearming at the Japanese port of Sasebo, *New Jersey* once again bombarded Wonsan and Kansong between June 4 and 7, this time with Adm. Arthur W. Radford, commander in chief of the Pacific Fleet, and Vice Adm. C. Turner Joy, commander of Naval Forces Far East, embarked. On the 7th *New Jersey* replenished from the store ship *Graffias* (AF 29) and refueled from the oiler *Ashtabula* (AO 51).

On the 12th, with Task Force 77 steaming sixty miles off the Korean coast, the destroyer *Walke* (DD 723), then

A 16-inch shell photographed on exiting the barrel.
USN

New Jersey shelling the Korean coast in 1951.
USN

New Jersey on returning from her first deployment to Korea.
USN

positioned to starboard of *New Jersey*, struck a mine that claimed twenty-six lives and left her severely damaged. The bodies of her dead were transferred to *New Jersey* as the only vessel in company with facilities to act as a morgue. With this, *New Jersey* turned for Yokosuka, which she reached on the 15th, but she was soon back in action in support of UN troops advancing in the vicinity of Kansong on July 6. While proceeding at speed on the 16th, however, boilers 7 and 8 had to be shut down in fireroom 4 owing to a feed water problem. The defect was traced to a lubrication pump in engine room 4 that caused a chain reaction leading to the immobilization of propeller shaft 3, requiring *New Jersey* to return to Yokosuka for repairs, where she spent three weeks alongside the repair ship *Ajax* (AR 6).

The bombardment missions resumed on August 17 and continued until September 23. On October 1, Gen. Omar N. Bradley and Gen. Matthew B. Ridgway came on board to confer with Vice Admiral Martin. Four days later *New Jersey* was steaming off Hungnam and Hamhung when she was requested to deploy her helicopter to rescue a pilot from the carrier *Bonhomme Richard* (CV 31) who had come down near a river ten miles south of Wonsan. Flying under air cover, the helicopter rescued the wounded aviator and brought him back to *New Jersey* for treatment in the ship's hospital.

Early November found her shelling an assortment of targets in Iwon, Chongjin, Tanchon, and Kansong, including bridges, tunnels, command positions, artillery pieces, and railway lines. Assisted by aircraft from the Australian light carrier HMAS *Sydney*, on the 13th *New Jersey* shelled the Changsan-Got peninsula.

On the 21st she was relieved by *Wisconsin* to which Vice Admiral Martin now shifted his flag.

Stopping at Sasebo, *New Jersey* then proceeded to Hawaii escorted by the heavy cruisers *Helena* (CA 75) and *Toledo* (CA 133), reaching Long Beach on December 8 and finally Norfolk on the 20th via the Panama Canal. The holiday season provided an opportunity to turn over most of the crew, and on January 4, 1952, Rear Adm. H. Raymond Thurber and his staff came on board *New Jersey*, which proceeded to Portsmouth Naval Shipyard on February 11 for updates to her electronic systems. After spending spring and summer at Norfolk and off Cuba, *New Jersey* was in readiness in early July to receive no less than 731 midshipmen from Annapolis and the NROTC of a dozen colleges who embarked on the 4th. On the 19th *New Jersey* sailed for Europe with the light cruiser *Roanoke* (CL 145), the fleet oiler *Severn* (AO 61), and seven destroyers. First stop was the French port of Cherbourg, where the famous Cunard liner RMS *Queen Mary* tied up astern of *New Jersey* on August 2. This was followed by visits to Lisbon before the squadron returned to the Caribbean and finally Norfolk on September 4. The rest of the year was divided between Chesapeake Bay with reserve officers in training and on maneuvers with the carrier *Coral Sea* (CV 43), with the new year being seen in at Hampton Roads.

On March 5 *New Jersey* sailed from Hampton Roads, and after traversing the Panama Canal on the 9th called briefly at Long Beach, Pearl Harbor, and Yokosuka before resuming station off Korea, where she relieved *Missouri* in early April after embarking the commander in chief of the Seventh Fleet, Vice Adm. Joseph J. Clark. By the 13th she was back on the gun line and early that morning fired 104 16-inch shells against the Songjin area. During this action the destroyer *Laws* (DD 558) reported spotting a mine to port of *New Jersey* that was promptly dealt with with light guns as the bombardment continued. On the 15th the target was Kojo followed two days later by Hungnam and then Songjin again.

On May 14, *New Jersey* sailed for Inchon, where South Korean president Syngman Rhee was received on board, after which the crew was able to enjoy a few days of leave in Yokosuka. A spell at Sasebo followed in early June, and on the 16th the ship's helicopter (nicknamed "The Jersey Bounce") rescued the pilot of an F4U-5 Corsair from the carrier *Philippine Sea* (CV 47) that had ditched near *New Jersey* after developing engine trouble.

Bombardment missions continued, often in the company of the heavy cruiser *Saint Paul* (CL 73), until July 28, with enemy positions being targeted at Songjin, Hodo Pando Island, Kalmagak, Wonsan, Chinampo, Kosong, Hungnam, and Tanchon. Hostilities ended on the 27th, however, and in late August the crew was able to enjoy a week's leave in Hong Kong. After a long wait, *New Jersey* was finally relieved by *Wisconsin* on October 14, and two days later sailed for home, reaching Norfolk via Pearl Harbor, Long Beach, and the Panama Canal (in which she suffered a slight collision in the Miraflores locks) on November 14.

During her second deployment to Korea *New Jersey* had sailed 37,519 miles and fired more than four thousand 16-inch shells, claiming the following successes in the process:

	ARTILLERY POSITIONS	BUILDINGS	TUNNELS	HIGHWAY BRIDGES	RAILROAD BRIDGES	BUNKERS	OBSERVATION POSTS	COMMON POSTS
Damaged	124	37	—	3	7	182	4	2
Destroyed	144	75	18	3	3	255	8	1

Source: Muir, *The Iowa Class Battleships*, 150.

New Jersey opens up with her main battery against Kaesong, Korea, on January 1, 1953. Note the parabolic array for the SPS-6 radar above the main rangefinder.
USN

A Sikorsky HO3S-1 landing on *New Jersey*'s fantail in April 1953.
USN

New Jersey after recommissioning, April 1968.
NARA

New Jersey in the Panama Canal on June 4, 1968.
USN

New Jersey prepares to carry out a 16-inch shoot. USN

Shortly after 0700 on the 30th, the crew was called to general quarters and turret 2 fired its first salvos against targets north of the Benhai River. Each impact dug a crater twenty feet deep and fifty feet wide with a two-hundred-yard bursting diameter and that cleared tropical vegetation over a diameter of three hundred yards. The exercise was repeated on October 1 with replenishment from the ammunition ship *Haleakala* (AE 25) the following day.

On the morning of October 7, small enemy vessels were spotted proceeding along the North Vietnamese coast in the Song Giang River area by a Grumman S-2 antisubmarine aircraft. *New Jersey* and *Towers* took them under fire with their 5-inch guns, destroying seven and forcing the rest ashore, and *New Jersey* spent the rest of the month pounding enemy positions with spotting for fall of shot provided by Vought A-7 Corsair II aircraft from the carrier *America* (CVA 66). On November 1 she embarked another 417 tons of shells and four days later carried out eight fire support missions for the 173rd Airborne Brigade, destroying command and fortified positions. The 25th was the busiest day of the war for *New Jersey*, which bombarded eight different sites destroying 117 buildings and 32 command positions before proceeding to the Huế area to support the 101st Airborne Division.

New Jersey in action against Vietnamese positions. USN

On December 8 she participated in the Meade River Operation, bombarding command positions south of Đà Nẵng. The holiday season was marked by the arrival by helicopter of Bob Hope, Ann-Margret, and a troupe of nineteen dancers who put on a show on the roof of turret 1. *New Jersey* was soon back in action, however, participating in Operation Bold Mariner, a landing on the Batangan peninsula between January 11 and 13, 1969.

On February 14, *New Jersey* shelled an artillery position and the following day silenced rocket launchers in the Cồn Tiên area. On the 22nd she was required to intervene urgently in the Demilitarized Zone (DMZ) to relieve U.S. Army units in the face of a Viet Cong attack, a mission that took her close to the 17th parallel that demarcated North and South Vietnam. After returning to Subic Bay on March 13, *New Jersey* was again in action off Cam Ranh Bay from the 21st. Seven days later she was cruising between Phan Thiết and Tuy Hòa, and from then until April 1 *New Jersey* provided support to the 3rd Marine Division from a position just south of the DMZ.

On April 3, *New Jersey* quit the China Sea after 120 days in action during which 5,866 16-inch and 14,891 5-inch shells were fired. After three days at Yokosuka, on April 9 she sailed for Long Beach with the carrier *Coral Sea* and three destroyers. These orders, however, were premature and on the 15th, with the ship 1,800 miles from her destination, she was ordered to turn back and resume station off Vietnam, the result of the shooting down of a Lockheed EC-121 by a North Korean aircraft. The feelings of the crew are to be imagined. *New Jersey* duly steered at 22.4 knots for Yokosuka, which she reached on the 22nd and immediately embarked 837 tons of ammunition from the *Paricutin* (AE 18). Although the ship was ready for anything, four days later a signal was received in the ship that Lt. Randy Ghilarducci translated as follows for the benefit of the crew: "Right full rudder, all engines ahead full, indicate turns for 22 knots, steer course 90." *New Jersey* reached Long Beach after an uneventful passage on May 5 to find more than a thousand people waiting to greet her at the pier. The ship was in considerable need of maintenance, and on the completion of repairs Rear Adm. Lloyd Vasey hoisted his flag on board on June 2. Curiously, the refitting had not extended to the admiral's quarters, which Vasey found dilapidated. Nonetheless, a summer training cruise began for 104 midshipmen on the 9th, with *New Jersey* joining Task Group 10.1 two weeks later. Visits followed to San Francisco, where ten thousand visitors were received on board, and then to Tacoma, Pearl Harbor, and San Diego before the ship returned to Long Beach on July 31, though not before the old minesweeper *Raven* (AM 55) had been expended as a gunnery target.

It was anticipated that *New Jersey* would sail to resume her place on the gun line off Vietnam on September 5, but three days earlier news was received from the Department of Defense as the ship was taking on ammunition at Seal Beach that she would in fact be withdrawn from active service at the earliest opportunity. On the 8th *New Jersey* sailed from Long Beach for Bremerton, where she tied up near *Missouri* and the cruisers *Quincy*, *Pasadena*, and *Pittsburgh*. Eight days later her colors were hauled down on a rainy and overcast day.

New Jersey during her deployment off Vietnam.
USN

New Jersey and the carrier *Coral Sea* in 1969.
USN

New Jersey enters San Francisco Bay in July 1969.
USN

New Jersey and *Missouri* at Puget Sound Naval Shipyard.
USN

The Happy Ship

From 1972 opinions were voiced concerning the conversion of *Missouri* to a museum, maintenance of *New Jersey* in reserve, and the scrapping of *Iowa* and *Wisconsin*. Despite these suggestions, which were the subject of constant debate in the U.S. Navy, several senior officers favored keeping these four large ships on standby. The truth is that the *Iowa* class was saved from destruction by their formidable 16-inch guns, designed decades earlier but with no equivalent in any other branch of the military.

In 1975 the Battleship New Jersey Museum came into being in the expectation of seeing the ship converted to a museum and berthed permanently at Sandy Hook near New York, but as it turned out "The Big J" had further years of service left in her.

In 1981 the determination of the Reagan administration to build a six-hundred-ship Navy and the need to respond to the Soviet *Kirov*-class cruisers prompted Congress to allocate $421 million to modernize *New Jersey*, with similar plans for her sisters to strengthen the U.S. presence in the Red Sea and the Caribbean. It remained to find out what solutions would be proposed to modernize them.

On July 27, 1981, *New Jersey* was towed to the Long Beach Naval Shipyard for refitting. Four of the after 5-inch mounts were landed together with the mainmast and the after crane. In their place came thirty-two Tomahawk cruise missiles, sixteen Harpoon missiles, four Phalanx close-in weapon system (CIWS) guns, and a complete radar fit. By March 1982 the modernization was in the process of completion and well ahead of that of her sisters, since, unlike them, most of her obsolete equipment had already been removed when she was commissioned for service off Vietnam.

In September *New Jersey* ran her first steam trials off the island of San Clemente near San Diego, and on October 18 carried out exercises with her 16-inch, 5-inch, and Phalanx weaponry with Secretary of the Navy John Lehman embarked. In mid-November the Board of Inspection and Survey (InSurv), the body responsible for inspecting and assessing the material condition of U.S. Navy vessels, approved *New Jersey*'s acceptance to the fleet.

The commissioning ceremony held on December 28 was attended by President Reagan and more than ten thousand others. On March 7, 1983, *New Jersey* sailed to carry out exercises off Southern California, and on the 23rd she became the first battleship to fire a Harpoon missile, destroying a decommissioned landing craft at a range of forty miles. After further exercises and a visit to San Francisco in April, *New Jersey* joined the Third Fleet to participate in ReadiEx 1-83, a simulated engagement off the West Coast during which the guided-missile destroyer *Callaghan* (DDG 994) and the guided-missile frigate *John A. Moore* (FFG 19) refueled from *New Jersey*.

It remained to prove to her critics that *New Jersey* could launch an attack over the horizon, and May 10 found her cruising off San Nicolas Island with numerous experts and technicians embarked. Early that afternoon a Tomahawk cruise missile was fired at a target five hundred miles away in Tonopah, Nevada. The trials were a success and the vessel returned to Long Beach two days later.

New Jersey drydocked in 1981.
USN

New Jersey towed to Long Beach for refitting prior to recommissioning.
USN

A technician works on one of the starboard propellers.
USN

President Reagan delivers a speech during *New Jersey*'s recommissioning ceremony on December 28, 1982.
USN

New Jersey firing a Tomahawk cruise missile off California on May 10, 1983.
USN

New Jersey entering Manila Bay with Fort Drum in the foreground. USN

On June 20, *New Jersey* sailed from Pearl Harbor to the Philippines, anchoring off Manila on July 3. Between the 6th and the 11th she participated in the Battle Week 2-83 Exercise focused on the carrier *Midway* (CV 41), during which she accounted for another decommissioned warship with a Harpoon missile. Visits followed to Singapore and Pattaya Beach in late July, but scheduled stops at Hong Kong, Guam, South Korea, and Japan were canceled after *New Jersey* was ordered to sail for Central America to consolidate the U.S. naval presence in the region. Soviet vessels of all types had for some time been arriving at Nicaragua with shipments of what were suspected to be weapons for the Sandinista regime. After resupplying at Luzon, *New Jersey* sailed for Pearl Harbor with the guided-missile cruiser *Leahy* (CG 16), the destroyers *Robison* (DDG 12), *Buchanan* (DDG 14), and *Ingersoll* (DD 990), and the frigate *Roark* (FF 1053), and she took the opportunity to carry out exercises with the carrier *Ranger* (CV 61) between August 11 and 14. Although *New Jersey* arrived on station on the 26th, the assignment was uneventful, and on September 15 she received a visit from Secretary of Defense Caspar Weinberger and Salvadoran president Álvaro Magaña at Balboa at the Pacific entrance to the Panama Canal.

New Jersey in 1984.
PRADIGNAC AND LEO

Meanwhile, severe unrest in the Middle East with heavy casualties to U.S. forces in Beirut from the spring of 1983 prompted the Reagan administration to reinforce its presence in the eastern Mediterranean, notably by assigning *New Jersey* to the Sixth Fleet. After passing through the Panama Canal, *New Jersey* called at the Spanish port of Rota on September 21 and appeared off the Middle Eastern coast on October 3. The conflict was escalated on the 23rd by the Beirut barracks bombing, which claimed more than three hundred lives. On December 14, *New Jersey* fired eleven 16-inch shells at Syrian and Druze antiaircraft positions, the first time she had opened fire in anger since Vietnam in 1969. The ship, which took on a new crew in early December, was entertained by Bob Hope and a troupe of celebrities on Christmas Eve and saw in the new year at the Israeli port of Haifa. She then resumed patrol duty, which usually consisted of steaming at between 5 and 10 knots three miles off the Lebanese coast with turret 1 kept at constant readiness. On February 7, 1984, *New Jersey* fired thirty-two 5-inch shells against enemy positions, but this was insignificant by comparison with the bombardment that came the following day when 288 16-inch shells were unleashed against Druze and Shi'ite positions near the town of Hammana sixteen miles east of Beiruit in the greatest naval bombardment since the Korean War. On the 26th another thirty 16-inch shells were fired in support of the deployment of the 22nd Marine Division. After these exertions, *New Jersey* put in at Haifa in early March and on April 2 Capt. Richard D. Milligan was able to inform his crew that they were leaving the combat zone. Stops at Naples and Villefranche and a traverse of the Panama Canal brought her to Pier 1 at Long Beach on May 5 where a crowd of five thousand awaited. Morale had been excellent throughout the deployment, and *New Jersey* earned the sobriquet of "The Happy Ship" from her crew.

New Jersey in 1984.
PRADIGNAC AND LEO

New Jersey engaging Syrian artillery positions overlooking Beirut.
USN

Another view of *New Jersey* engaging Syrian artillery positions near Beirut.
USN

New Jersey fires a 16-inch salvo on either beam.
USN

Turret 2 in action.
NARA

New Jersey anchored at Villefranche in the south of France.
PRADIGNAC AND LEO

On June 15, *New Jersey* was transferred to the naval shipyard to prepare for replacement of the central gun of turret 2, a task completed two months later thanks to the ex-German floating crane *Titan II*. By December 3 she was rearming at Seal Beach Weapon Station, and on February 4, 1985, she sailed for San Diego to carry out gunnery exercises off San Clemente in mid-April. *New Jersey* then returned to Long Beach for an engine refit that lasted until late June.

On July 10 she sailed for Pearl Harbor with the guided-missile cruisers *England* (CG 22), *Halsey* (CG 23), and *Jouett* (CG 29), the destroyers *Chandler* (DDG 996), *O'Brien* (DD 975), and *Ingersoll* (DD 990), and the frigate *Reasoner* (FF 1063). A week of training exercises brought her to San Francisco on August 8, where she received 21,000 visitors over the next few days.

The following year, 1986, was punctuated by a series of major exercises, beginning with Computex 86-2 on January 21, during which she fired missiles; Readiex 86-2 with the guided-missile cruiser *Long Beach* (CGN 9), the destroyers *Merrill* (DD 976) and *Fletcher* (DD 992), the frigates *Bronstein* (FF 1037), *Stein* (FF 1065), *Copeland* (FFG 25), and *Thach* (FFG 43), and the replenishment oiler *Wabash* (AOR 5). Then on May 28 came the start of exercise RimPac 86 carried out in the Pacific with units of the Royal Navy, to be followed on August 5 with Cobra Gold 86 in the Gulf of Thailand. *New Jersey* returned to Long Beach Naval Shipyard on December 11. After a quiet year in 1987, she returned to Korean waters in the summer of 1988 to participate in RimPac 88 before heading for Australia to join in that country's bicentennial celebrations.

On May 19, 1989, *New Jersey* received her final commanding officer, Capt. Ronald Tucker, who was given the sad duty of deactivating the vessel. Her last mission of note was PACex 89 held with Canadian, Korean, and Japanese marines in the northern Pacific between September and October of that year and involving fifty-five ships of all types.

New Jersey subsequently sailed for the Indian Ocean and ended the year in the Persian Gulf before returning to the United States in February 1990, a circumnavigation that brought a fitting end to a remarkable career.

New Jersey at Villefranche in April 1984.
PHILIPPE CARESSE

New Jersey sailing from Villefranche.
PRADIGNAC AND LEO

New Jersey traversing the Panama Canal.
USN

New Jersey leading *Missouri* and the guided-missile cruiser *Long Beach* (CGN 9) in July 1988.
USN

A gunnery exercise by *New Jersey* off Sydney in October 1988.
NARA

The carrier *Enterprise* (CVN 65) and *New Jersey* during the fleet exercises of October 1989.
USN

Camden's Battleship

In 1990 it was decided to keep *New Jersey* and *Wisconsin* in the reserve fleet while *Iowa* and *Missouri* were released for preservation as museums. Budget cuts, however, resulted in *New Jersey* being deactivated for the last time at Long Beach on February 8, 1991, and the ship subsequently being towed to Bremerton. Meanwhile, the Battleship New Jersey Museum began its search for the most appropriate berth for the vessel. After considering mooring her at Liberty State Park in Jersey City, it was decided to bring her to Port Alliance at Camden on the Delaware River due to its historical significance. Funding was provided by the Camden Empowerment Zone Corporation, which donated $1 million, the City of Camden, which gave $3.2 million, and the governor of New Jersey, Christie Whitman, who committed $6 million. It remained, however, to transfer the vessel to the East Coast, for which purpose the State of New Jersey immediately released $2 million in view of the $300,000 cost of traversing the Panama Canal. *New Jersey* finally entered the Delaware on November 11, 1991, to the cheers of 25,000 spectators.

New Jersey at Pearl Harbor in August 1990.
USN

New Jersey towed into Philadelphia Naval Shipyard on November 11, 1991.
USN

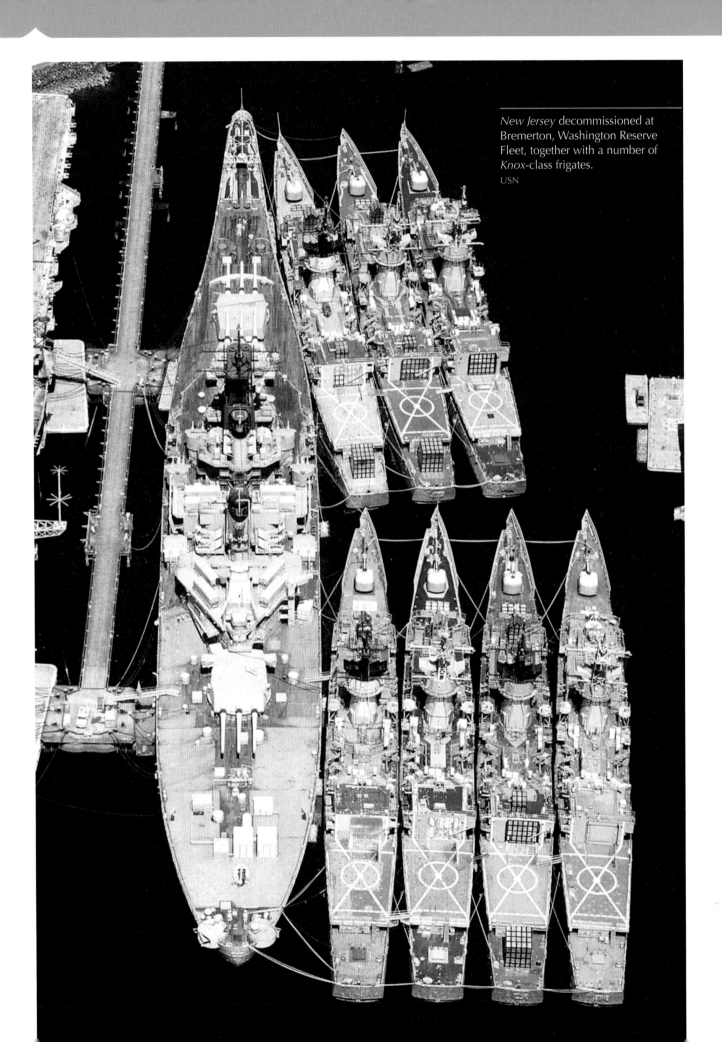

New Jersey decommissioned at Bremerton, Washington Reserve Fleet, together with a number of *Knox*-class frigates.
USN

On August 15, 2000, the battleship was towed to the South Jersey Port Corporation for refitting as a museum. After fourteen months of work during which a hundred volunteers gave 90,000 hours of labor, the attraction opened its doors on October 15, 2001. The pier and infrastructure at Port Alliance had cost $12 million.

A walk along Clinton Street on the banks of the Delaware reveals the vessel in all her glory. Stepping on board, the imposing bulk of turrets 1 and 2 dominate the scene. Descending through a hatch on the forecastle, it is possible to view the anchor windless, capstan machinery compartment, and crew quarters. Reemerging beside turret 1, the visitor enters the forward superstructure containing, among other spaces, the officers' and admiral's quarters before ascending to the admiral's bridge, signal bridge, and the navigation bridge with its equipment intact and the chart room aft. Visible from the forward superstructure are four of the eight armored box launchers (ABLs) housing the Tomahawk cruise missiles, one of which can be seen emerging from its housing, together with the Phalanx CIWS guns and the SRBOC missile decoy launchers. A Kaman SH-2F Seasprite is preserved on the fantail and turret 3 can be visited. Other highlights are the radio room, the ship's laundry, brig, bakery, and barbershop, and the tour is complemented by extensive exhibitions and displays of ship models.

> Battleship *New Jersey* Museum and Memorial
> 62 Battleship Place
> Camden Waterfront
> Camden, NJ 08103 U.S.A.
> www.battleshipnewjersey.org

USS *NEW JERSEY* (BB 62)

Commissioned	May 23, 1943
Decommissioned	June 30, 1948
Commissioned	November 21, 1950
Decommissioned	August 21, 1957
Commissioned	April 6, 1968
Decommissioned	December 17, 1969
Commissioned	December 28, 1982
Decommissioned	February 8, 1991

New Jersey at her berth at Camden, New Jersey, May 1933.
PHILIPPE CARESSE

An enlisted men's canteen.
PHILIPPE CARESSE

Enlisted men's quarters in *New Jersey*.
PHILIPPE CARESSE

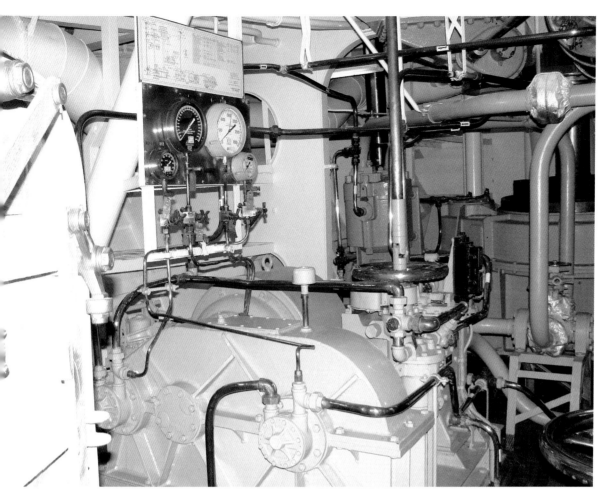

The anchor windless and capstan machinery compartment.
PHILIPPE CARESSE

The interior of turret 3.
PHILIPPE CARESSE

The forecastle seen from the bridge.
PHILIPPE CARESSE

The upper conning tower with its massive armored door.
PHILIPPE CARESSE

Turret 3 trained to starboard.
PHILIPPE CARESSE

Interior of a 5-inch mount.
PHILIPPE CARESSE

Fitted abaft the after stack is the Mk-13 radar with its Mk-38 rangefinder. In the foreground is the Mk-37 rangefinder surmounted by the Mk-25 radar.
PHILIPPE CARESSE

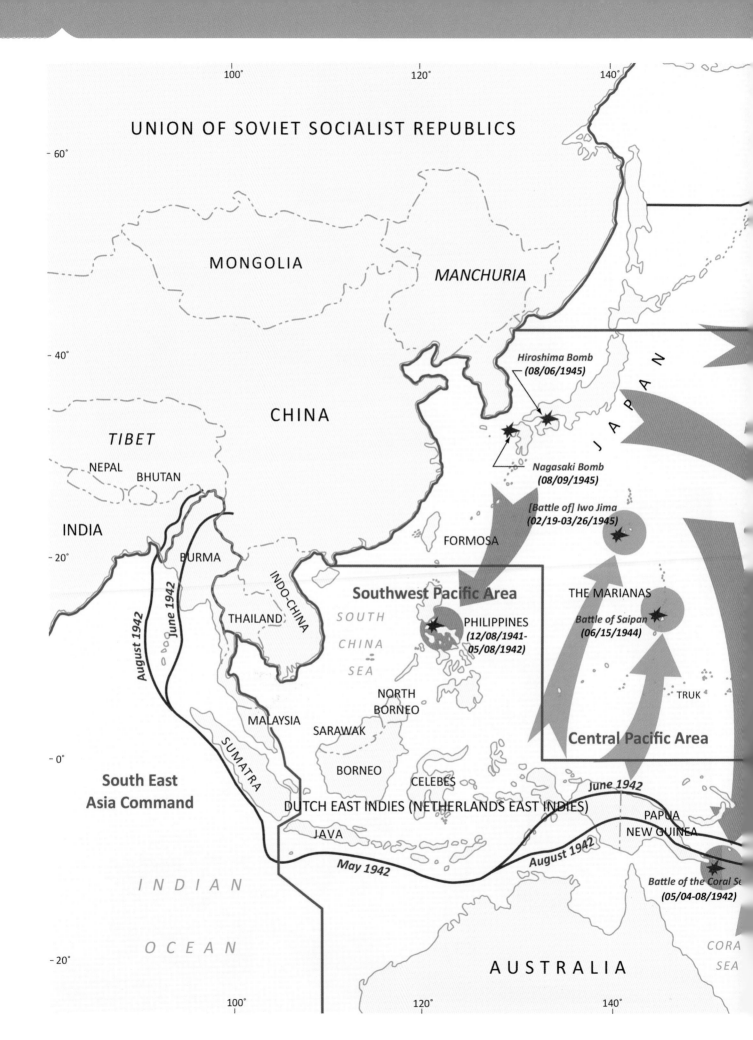

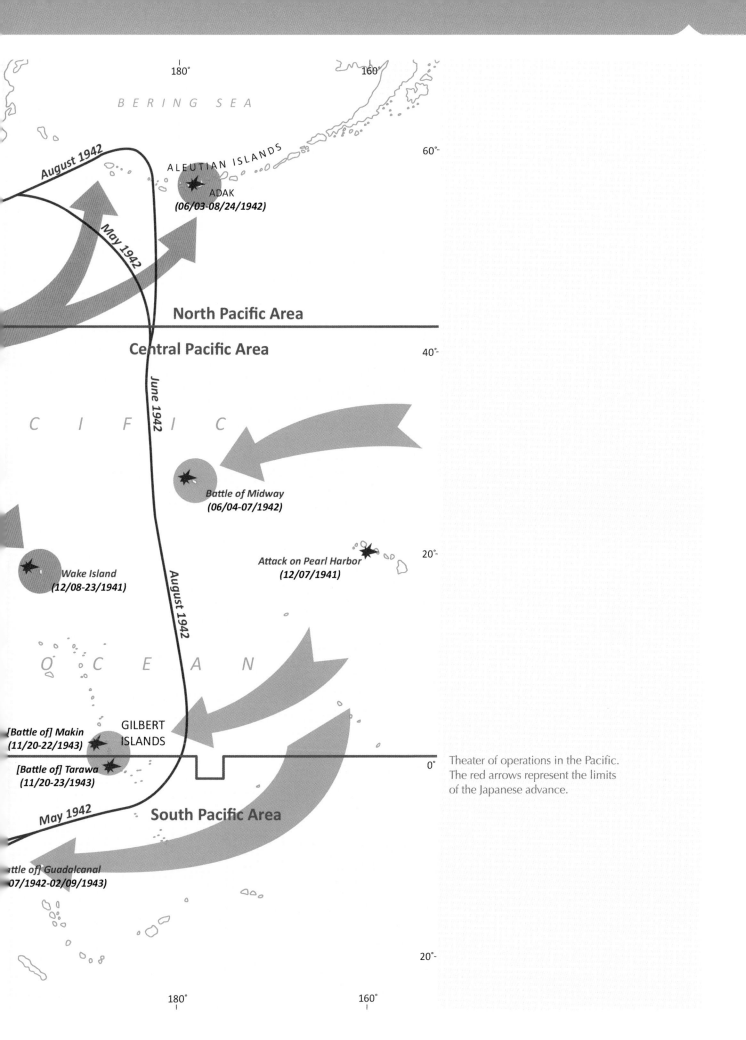

Theater of operations in the Pacific. The red arrows represent the limits of the Japanese advance.

Theater of operations in Korea

Theater of operations in Vietnam

FIFTEEN

The Career of the Battleship USS *Missouri* (BB 63)

The Last Battleship

The keel of the USS *Missouri* (BB 63) was laid at New York Naval Shipyard on January 6, 1941, and the ship was launched on January 29, 1944. Her sponsor on that occasion was Margaret Truman, daughter of the junior senator from Missouri and future president of the United States, Harry S. Truman, and it was the beginning of a close relationship with the ship and her crew. Among those present was Adm. William F. Halsey, who made this brief speech to those present: "We have a date to keep in Tokyo. Ships like the *Missouri* will provide the wallop to flatten Tojo and his crew." Fifteen minutes later *Missouri* entered the water.

The work of completion lasted forty-one months and *Missouri* was commissioned on June 11, 1944, under the command of Capt. William Callaghan in a ceremony presided over by Secretary of the Navy James Forrestal and Senator Truman and Senator Joel B. Clark. She was destined to be the last battleship to enter service with the U.S. Navy. After being brought to full complement at Bayonne, *Missouri* headed for Chesapeake Bay on August 3 for trials that lasted until the 21st followed by speed, gunnery, and catapult trials in the Gulf of Paria off Venezuela. An early casualty of *Missouri*'s service was the sailor crushed by the traversing of a 40-mm mount.

Laying *Missouri*'s triple bottom on October 3, 1941.
USN

Missouri's hull under construction in 1943.
USN

Missouri ready for launching.
NARA

Missouri on the eve of her launch.
NARA

Missouri enters the water on January 29, 1944.
USNHC

Margaret Truman in the act of launching *Missouri*. Her father, Missouri senator and future president Harry Truman, is behind her.
USN

Missouri in dry dock in New York. Camouflage Ms 32/22D has already been applied.
USN

Missouri in the final stages of completion at New York Naval Shipyard.
USN

Missouri in the final stages of completion at New York Naval Shipyard.
USN

An Mk-38 fire-control director being hoisted into place.
USN

Missouri commissioned on June 11, 1944.
NARA

Missouri proceeds to sea for the first time with the assistance of tugs.
USNHC

Missouri during her shakedown cruise in July 1944.
NARA

Missouri during her shakedown cruise in July 1944.
NARA

Missouri running trials in the Chesapeake Bay.
NARA

By early November *Missouri* was ready to join the war in the Pacific, and on November 10 she sailed from New York with Task Group 27.7, reaching San Francisco via the Panama Canal on the 18th. After completing the final preparations, "The Mighty Mo" or "Big Mo," as she had by now been christened, sailed for Pearl Harbor on December 8 escorted by the destroyers *Bailey* (DD 492) and *Terry* (DD 513). After a week at Pearl Harbor, *Missouri* sailed for the anchorage at Ulithi in the Caroline Islands on January 1, 1945, reaching it on the 13th, where she joined Adm. Marc A. Mitscher's Task Force 58, itself part of Adm. Raymond Spruance's Fifth Fleet. Assigned to Task Group 58.2 under Rear Adm. Ralph E. Davison, *Missouri* joined her sister ship *Wisconsin* together with the carriers *Lexington* (CV 16) and *Hancock* (CV 19), the escort carrier *San Jacinto* (CVL 30), the heavy cruisers *Boston* (CA 69) and *Pittsburgh* (CA 72), and thirteen destroyers. TG 58.2 was reinforced in February 1945 by the carrier *Enterprise* (CV 6), the large cruiser *Alaska* (CB 1), the heavy cruisers *San Francisco* (CA 38) and *Baltimore* (CA 68), the light cruiser *Flint* (CL 97), and seven destroyers. *Missouri* remained in the comparative safety of this anchorage until February 10, when she sailed with TG 58.2 to support the ongoing assault on Iwo Jima.

While preparing to shell the Ogasawara archipelago on the night of the 19th, *Missouri*'s radar detected the approach of unidentified aircraft, and the resulting action secured the ship her first kill in the shape of a Nakajima Ki-49 Helen bomber. *Missouri* remained in a support role for the Iwo Jima operation in a position sixty-five miles northwest of the island until the 24th, after which she and TG 58.2 sailed to rendezvous with the oiler two hundred miles east of Iōtō, its progress delayed by a violent storm that obliged it to reduce speed to 16 knots. *Missouri* suffered damage and electrical problems in four of her 40-mm mounts as a consequence of a forty-foot swell that also damaged the sights of the forward 16-inch turret. Task Force 58 subsequently proceeded to a point 190 miles southeast of Tokyo to launch raids on that city between the 24th and 26th. Attacks on Okinawa followed on March 1. The return voyage to Ulithi was uneventful except for the loss of one of *Guam*'s SC-1 Seahawk floatplanes during an attempted transfer to *Missouri*, the wreck having to be sunk by gunfire from a destroyer. By the time *Missouri* reached Ulithi on March 4 she had steamed more than eight thousand miles on her maiden war cruise.

Missouri during her trials in the Chesapeake Bay.
NARA

Missouri fires her main battery in August 1944.
USNHC

at 0805 *Missouri* was attacked by Helen bombers and Kawasaki Ki-45 Nick fighters, replying with a heavy AA barrage that claimed several aircraft. The attacks continued, however, and at 1316 *Missouri* destroyed an aircraft that had just released a bomb against *Yorktown* and followed this up by downing a second aircraft minutes later; not until 2115 did the last Japanese attack subside.

On the 19th, U.S. air raids were launched against air bases in the Ryūkyū Islands as well as against Kōbe and Kure, causing damage to heavy units of the Imperial Japanese Navy. The counterattack proved lethal, however, as the carrier *Wasp* (CV 18) of Task Group 58.1 suffered heavy casualties to a bomb hit and Rear Adm. Ralph E. Davison's flagship, the carrier *Franklin* (CV 13) of Task Group 58.2, was set ablaze by two bombs. Nonetheless, Admiral Mitscher kept up the pressure, and the three battleships and their escort bombarded Kutaka Shima on the 24th after refueling six hundred miles east of Ryūkyū. That afternoon it was the southeast coast of Okinawa that came under naval bombardment.

Missouri exercising with the large cruiser USS *Alaska* (CB 1).
NARA

An OS2U Kingfisher floatplane being hoisted onto its catapult.
NARA

Refueled, replenished, and rearmed, on March 13 *Missouri* was reassigned to Task Group 58.4 under Rear Adm. Arthur W. Radford. TG 58.4 consisted of, in addition to *New Jersey* and *Wisconsin*, the carriers *Enterprise* (CV 6), *Intrepid* (CV 11), and *Yorktown* (CV 10), the large cruisers *Alaska* (CB 1) and *Guam* (CB 2), the light cruisers *San Diego* (CL 53), *Oakland* (CL 95), and *Flint* (CL 97), together with twenty-four destroyers. The following day TG 58.4 sailed to participate in Operation Iceberg, the invasion of Okinawa. On the morning of the 18th the fleet was attacked by Japanese aircraft east of Kyūshū and

The landings on Iwo Jima, February 19, 1945.
USN

An antiaircraft armament exercise during the Pacific War.
USN

Missouri and her escort under Japanese air attack.
USN

Between March 24 and April 19, a bombardment group was established as Task Group 58.7 under Rear Adm. John F. Shafroth. It included *New Jersey*, *Missouri*, *Wisconsin*, *North Carolina* (BB 55), *Washington* (BB 56), *South Dakota* (BB 57), *Massachusetts* (BB 59), and *Indiana* (BB 58). This formidable force was given the task of shelling installations at locations planned for troop landings, with *Iowa*'s floatplanes spotting fall of shot. The invasion of Okinawa began on April 1, and five days later a B-29 reported that the battleship *Yamato*, a light cruiser, and eight destroyers were under way in the Bungo Strait. This was the start of Ten-Ichi-Gō, a last throw of the dice for the Imperial Japanese Navy, which aimed to put *Yamato* ashore as a floating battery against the U.S. invasion forces. But it was not to be, as *Yamato* was destroyed on the 7th by eight bombs and fourteen torpedoes delivered by aircraft from Task Force 58.

After *Missouri* shelled airfields in southern Ryūkyū on the morning of the 11th, a report was received at 1330 of enemy aircraft approaching the fleet, and ten minutes later TG 58.7 was attacked by thirteen aircraft. *Missouri* opened a heavy barrage of fire but at 1442 a Zeke piloted by a kamikaze struck her starboard side at frame 169, causing damage to No. 17 40-mm mount and a fuel blaze abreast No. 53 5-inch mount. The flames were quickly brought under control and there were no U.S. casualties; the pilot received a military funeral the following day. Later that afternoon *Missouri* engaged and damaged two other aircraft at a range of 11,900 yards.

The famous photo taken at the moment a Zero fighter struck *Missouri* on the afternoon of April 11, 1945.
USN

One of the Zero's machine guns freakishly caught in the barrel of a 40-mm Bofors gun.
USN

The remains of the Zero on *Missouri*'s deck.
NARA

At 1303 on the 16th, another suicide attack materialized against the fleet, and at 1326 *Missouri* downed a Zero that disappeared near *Intrepid*. Two minutes later a second Zero was shot down, this time hitting the ship's stern crane and exploding over the fantail but without causing any casualties or severe damage. At 1335 yet another Zero attacked *Missouri* but was shot down 100 yards off the starboard bow, and *Missouri*'s crew, increasingly experienced in dealing with this threat, continued to repel attacks until 2111. *Missouri*'s luck held, but this was not true for others such as the carrier *Enterprise*, which was struck twice, *Essex* (CV 9), which suffered the loss of thirty-three men, and the destroyer *Kidd* (DD 661), which took a bomb in the hull that claimed the lives of thirty-eight men, with three more lost in the destroyer *Hank* (DD 702). On the 17th a submarine contact was established at a range of twelve miles, and *I-56* was sunk by aircraft from the carrier *Bataan* (CVL 29) and the destroyers of its escort. Later that day a Japanese squadron was intercepted and destroyed sixty miles from the fleet.

On May 5 *Missouri* turned for Ulithi, where she anchored on the 9th before sailing for Apra Harbor, Guam, on the 17th. At 1527 the following day, Admiral Halsey came on board with his staff of sixty officers and hoisted his flag in *Missouri*, which thereby became the flagship of the Third Fleet, renamed from the Fifth Fleet. On the 21st she sailed with the destroyers *McNair* (DD 679) and *Wedderburn* (DD 684) for Hagushi Bay in Okinawa at the head of Task Group 30.1. On the 27th the battleship sailed to participate in the bombardment of the southeast coast of Okinawa before rejoining Task Group 38.4 the following day. A slackening of enemy activity allowed raids to be carried out against airbases on Ryūkyū, but on June 4 a typhoon was reported to be approaching fifty miles south-southwest of Task Force 38, which began taking measures to distance itself from the threat. The plan was for *Missouri* to refuel from a number of oilers and a course of 100 degrees was laid, but at 0100 on the 5th the command ship *Ancon* (AGC 4) signaled Halsey that TF 38 was still in the path of the storm as a result of which a course alteration to 300 degrees was ordered. Despite these precautions Halsey's ships were assailed by 80-knot winds, and serious damage was done to the heavy cruiser *Pittsburgh* (CA 72), which lost her bow in hundred-foot waves, with further damage inflicted on the cruisers *Baltimore* (CA 68) and *Duluth* (CL 87). The maximum intensity of the storm was recorded by the escort carrier *Windham Bay* (CVE 92), which reported 127-knot winds and seventy-five-foot waves. By the time calm had been restored, TF 38 had lost thirty-three aircraft and six men with four battleships, four carriers, six cruisers, and ten destroyers requiring repairs with varying degrees of urgency.

A quiet moment on *Missouri*'s forward superstructure in 1945.
USN

Missouri unleashes a salvo against enemy positions from the forward turrets. The shells can be seen disappearing toward their target on the extreme right of the photo.
NARA

Missouri opens fire on enemy positions.
NARA

Missouri astern of a carrier during during a storm in the Pacific. Note that turret 1 has been trained to starboard.
USN

Between the 8th and the 10th, Ryūkyū was again a target of U.S. carrier aviation, after which it was the turn of Tokyo and the islands of Daitō-shotō. On the 10th the fleet sailed for San Pedro Bay in the Philippines for rest and replenishment, *Missouri* having steamed no less than 6,014 miles between May 28 and June 10. Once this respite ended on July 1, TF 38 sailed north to protect the carriers launching raids on Tokyo, lying off Honshū and Hokkaidō between the 11th and the 13th. Two days later she participated in a three-hour bombardment of the Nihon Steel and Wanishi Ironworks facilities at Muroran, with TF 38 (including *Iowa*) being reinforced by the battleships *North Carolina* (BB 55), *Alabama* (BB 60), and HMS *King George V*, the carrier HMS *Indefatigable*, and two British destroyers. During this action the center gun of *Missouri*'s turret 1 suffered from mechanical faults to the breech, requiring it to be temporarily brought out of action. Forty-eight hours later the ships were off Honshū shelling the installations at Hitachi. On the 18th *Missouri* rejoined Task Group 38.4 consisting of *Iowa*, *Wisconsin*, the carriers *Shangri La* (CV 38), *Yorktown* (CV 10), *Bonhomme Richard* (CV 31), *Essex* (CV 9), and *Cowpens* (CVL 25), four cruisers, and fourteen destroyers.

A huge replenishment operation after July 20 saw the transfer of 6,369 tons of ammunition, 1,635 tons of food, 63,600 tons of fuel, and 99 aircraft. On August 8 *Missouri* accounted for an Aichi B7A Grace dive bomber that came down within yards of the carrier *Wasp* (CV 18), and the following day resumed bombardment missions against the Japanese mainland. Meanwhile, the dropping of an atomic bomb first on Hiroshima and then on Nagasaki on August 6 and 9 prompted the Japanese government to accept the Allied surrender terms, and numerous senior officers came on board *Missouri* for conferences over the following days.

Missouri seen from *Iowa* during a jackstay transfer of personnel in August 1945. Someone has taken the trouble to fit tassels around the canopy of the chair.
USN

Missouri's forward 16-inch turrets.
NARA

The British battleship HMS *King George V* and *Missouri* in Japanese waters.
NARA

With the Japanese armed forces still reckoned to pose a major threat, TF 38 prepared to launch further strikes on Tokyo on the 14th, but no sooner had the planes taken off than they were recalled after CINCPAC announced the surrender of the Japanese Empire. At 0804, with *Missouri* steaming 120 miles southeast of Tokyo in position 34°00′ N, 142°11′ E, a message was received from President Truman confirming news of the surrender, and the aircraft had been recovered by 0840. At 1055 Adm. Chester Nimitz ordered a cessation of hostilities, and at 1110 the sirens of every vessel split the morning air. Nonetheless, the Third Fleet remained on high alert against the possibility of Japanese pilots failing to receive or accept the order to cease fire. *Missouri* continued steaming off the Japanese coast and on the 16th was visited by the commander in chief of the British Pacific Fleet, Admiral Sir Bruce Fraser, to confer with Halsey. On the 21st, Gen. Douglas MacArthur declared that the surrender would be signed on board a U.S. battleship, with *Washington* and *South Dakota* the obvious candidates in view of their distinguished war record. But President Truman had other ideas and selected the vessel that was not only named for his home state but that had been launched by his daughter.

On the 27th, *Missouri* and *Iowa* entered Sagami Bay twenty-five miles southwest of Tokyo. The destroyer *Hatsuzakura* had come alongside earlier with a civilian pilot and emissaries bearing maps and information on minefields that were received by Halsey's chief of staff, Rear Adm. Robert B. Carney. *Missouri* weighed anchor in the 28th and entered Tokyo Bay the following day accompanied by Admiral Fraser's flagship, HMS *Duke of York*, and units of the Third Fleet and British Pacific Fleet, all at general quarters. At 0925 she anchored off the naval base at Yokosuka to prepare for the long-awaited surrender ceremony. Nearby lay one of the few survivors of the Imperial Japanese Navy, the battleship *Nagato*. Over the course of the next few days the following vessels took their positions in Tokyo Bay:

On Saturday, September 2, Admiral Nimitz came on board *Missouri* at 0805, followed three quarters of an hour later by General MacArthur, who was to lead the ceremony. The Japanese delegation, conveyed by the destroyer *Lansdowne* (DD 486) and led by Foreign Minister Shigemitsu Mamoru and General Umezu Yoshijirō, arrived at 0856.

Also present were Admiral Sir Bruce Fraser representing the United Kingdom, General Hsu Yung-Ch'ang for China, Lieutenant General Kuzma Nikolaevich Derevyanko for the Soviet Union, General Sir Thomas Blamey for Australia, Colonel Lawrence Moore Cosgrave for Canada, General Philippe Leclerc de Hauteclocque for France, Vice Admiral Conrad Emil Lambert Helfrich for the Netherlands, and Air Vice Marshall Leonard M. Isitt for New Zealand, together with 225 war correspondents and 65 photographers strategically positioned to record the ceremony. At 0904 Foreign Minister Shigemitsu came forward to sign the Instrument of Surrender, and once the Japanese delegation and the representatives of the Allied powers had done so MacArthur uttered these final words: "Let us pray that peace be now restored to the world and that God will preserve it always. These proceedings are closed." Admiral Halsey then gave the word and at 0925 the sky over Tokyo Bay was darkened by 455 aircraft overflying the fleet toward Mount Fuji.

BATTLESHIPS
USS *New Mexico* (BB 40) USS *Mississippi* (BB 41) USS *Idaho* (BB 42)
USS *Colorado* (BB 45) USS *West Virginia* (BB 48) USS *South Dakota* (BB 57)
USS *Iowa* (BB 61) USS *Missouri* (BB 63) HMS *King George V*
HMS *Duke of York*

ESCORT CARRIERS
USS *Cowpens* (CVL 25) USS *Bataan* (CVL 29) HMS *Ruler*
HMS *Speaker*

HEAVY CRUISERS
USS *Boston* (CA 69) USS *Quincy* (CA 71) USS *St Paul* (CA 73)
USS *Chicago* (CA 136) HMAS *Shropshire*

LIGHT CRUISERS
USS *Detroit* (CL 8) USS *San Juan* (CL 54) USS *Pasadena* (CL 65)
USS *Springfield* (CL 66) USS *Oakland* (CL 95) USS *Wilkes Barre* (CL 103)
HMS *Newfoundland* HMAS *Hobart* HMNZS *Gambia*

DESTROYERS
USS *Mayo* (DD 422) USS *Madison* (DD 425) USS *Hilary P. Jones* (DD 427)
USS *Charles Hughes* (DD 428) USS *Nicolas* (DD 449) USS *Taylor* (DD 468)
USS *Buchanan* (DD 484) USS *Lansdowne* (DD 486) USS *Lardner* (DD 487)
USS *Twining* (DD 540) USS *Yarnall* (DD 541) USS *Wren* (DD 568)
USS *Kalk* (DD 611) USS *Caperton* (DD 650) USS *Cogswell* (DD 651)
USS *Ingersoll* (DD 652) USS *Knapp* (DD 653) USS *Colahan* (DD 658)
USS *Clarence Bronson* (DD 668) USS *Cotton* (DD 669) USS *Dortch* (DD 670)
USS *Gatling* (DD 671) USS *Healy* (DD 672) USS *Stockham* (DD 683)
USS *Wedderburn* (DD 684) USS *Halsey Powell* (DD 686) USS *Uhlmann* (DD 687)
USS *Wadleigh* (DD 689) USS *Ault* (DD 698) USS *Wallace Lind* (DD 703)
USS *De Haven* (DD 727) USS *Frank Knox* (DD 742) USS *Southerland* (DD 743)
USS *Blue* (DD 744) USS *Robert Huntington* (DD 781) USS *Benham* (DD 796)
USS *Cushing* (DD 797) USS *Perkins* (DD 877) HMS *Quality* (G 62)
HMS *Whelp* (R 37) HMS *Wizard* (R 72) HMS *Teazer* (R 23)
HMS *Tenacious* (R 45) HMS *Terpsichore* (R 33) HMS *Wager* (R 98)
HMAS *Napier* (G 97) HMAS *Warramunga* (I 44) HMAS *Nizam* (G 38)

Completing this mighty display of naval power were nine escort destroyers, two frigates, two sloops, thirty-one minesweepers, twelve submarines, twelve transports, and numerous landing craft and steamers of all types.

Missouri and *Iowa* entering Tokyo Bay.
USN

The Japanese delegation comes on board *Missouri* on September 2, 1945.
NARA

General Douglas MacArthur signs the Instrument of Surrender.
NARA

General Kuzma Derevyanko signs the Instrument of Surrender on behalf of the Soviet Union.
NARA

Foreign Minister Shigemitsu Mamoru signs the Instrument of Surrender.
NARA

The Japanese delegation leaves *Missouri*.
NARA

Missouri in Tokyo Bay from which she sailed for Guam on September 6.
NARA

The sky over Tokyo Bay blackened by a flypast of hundreds of U.S. carrier aircraft shortly after the signing of the Instrument of Surrender. Lying just ahead of *Missouri* is the light cruiser *Detroit* (CL 8).
USNHC

On September 6 *Missouri* sailed from Japanese waters for Apra Harbor, Guam, Admiral Halsey having shifted his flag to *South Dakota* the previous day. There she embarked troops for repatriation to the United States as part of Operation Magic Carpet, reaching New York in a thick fog on October 23 via Pearl Harbor and the Panama Canal. Moored at Pier 90 in the Hudson, she formed the focus of Navy Day and was visited by President Truman and his family, New York governor Thomas E. Dewey, and Fleet Adm. William D. Leahy, while also being opened to the public.

Missouri approaching Pearl Harbor on September 20, 1945.
USN

After refitting, *Missouri* sailed for exercises and training off Cuba and saw in the new year in home waters, but on March 22, 1946, she sailed from New York for a cruise of the Mediterranean as a demonstration against Soviet ambitions in the Balkans. Embarked in her for repatriation was the body of the long-serving Turkish ambassador to the United States, Münir Ertegün, who had died in November 1944. *Missouri* anchored off Istanbul on April 5, her officers being invited to dine on board the battlecruiser *Sultan Yavuz Selim*, formerly the *Goeben* of the Imperial German Navy. Further stops awaited at the Piraeus and Naples, where a special train was laid on to convey three parties of officers and men to Rome for an audience with Pope Pius XII. After a stop at Algiers, *Missouri* turned for home, reaching Norfolk with her escort on May 9.

The crew mustered on the forecastle in October 1945.
USNHC

Navy Day in the Hudson on October 27, 1945. Nearest the camera is the heavy cruiser *Augusta* (CA 31) followed by the carriers *Midway* (CV 41) and *Enterprise* (CV 6), the battleships *Missouri* and *New York* (BB 34), and the heavy cruisers *Helena* (CA 75) and *Macon* (CA 132).
USN

Three days later she sailed for Culebra Island to rejoin Admiral Mitscher's Eighth Fleet, returning to New York on the 27th after carrying out maneuvers.

In August *Missouri* sailed on another important foreign mission, this time to Brazil for the signing of the Treaty of Rio. The formalities completed, President Truman and his family came on board on September 7 together with numerous Secret Service agents and forty reporters with whom the ship sailed for Norfolk. The president occupied the captain's cabin while the First Lady and their daughter, Margaret, were accommodated in the admiral's quarters. The trip could hardly been more agreeable, with kind weather and the genial president spotted playing poker with the officers and joining the chow line with his family in the ship's canteen. After landing its guests on the 19th, *Missouri* turned for New York, where she was laid up for refitting until March 10, 1948.

Missouri in May 1946.
USN

Missouri lying at Istanbul between April 5 and 9, 1949, together with the destroyer *Power* (DD 839). To the right is the famous battlecruiser *Sultan Yavuz Selim*, formerly of the German navy.
USNHC

Emerging from dockyard hands, *Missouri* carried out exercises in Guantánamo Bay in preparation for receiving a contingent of 421 midshipmen from Annapolis. At 0430 on June 6, she sailed for England escorted by four destroyers and four minesweepers as part of Task Force 61, *Missouri* having by now earned the distinction of becoming the first battleship to embark and operate helicopters in the shape of two Sikorsky HO3S-1s. After an Atlantic crossing punctuated by numerous exercises and maneuvers, the squadron reached Portsmouth, where it spent eight days, during which some 32,000 visitors were received on board. After this agreeable interlude, TF 61 refueled from the oiler *Caloosahatchee* (AO 98) shortly before turning for the United States via the Azores and stopping in mid-ocean for some afternoon swimming for the crew. After a stop at Guantánamo Bay on July 7 the ship returned to Norfolk on the 21st.

President Truman admires the plaque indicating the precise spot where the Japanese Instrument of Surrender was signed.
USN

President Truman on board *Missouri* on the same occasion.
NARA

The plaque commemorating the Japanese surrender positioned on the starboard side of the forward superstructure.
USN

A meal being served cafeteria-style in the enlisted men's mess in *Missouri*. The importance of good food to morale can never be overestimated.
NARA

A petty officer enjoying the offerings of the ship's galley.
NARA

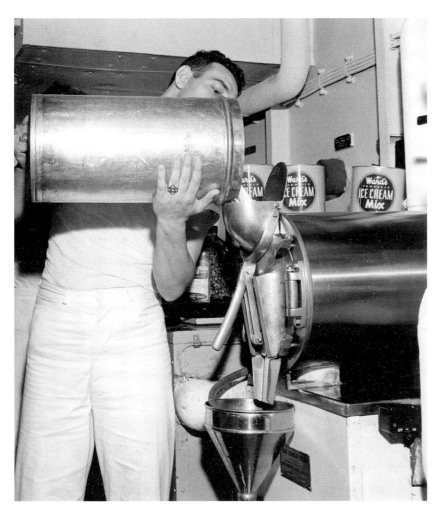

Ice cream in the course of preparation.
NARA

Payday in *Missouri*.
NARA

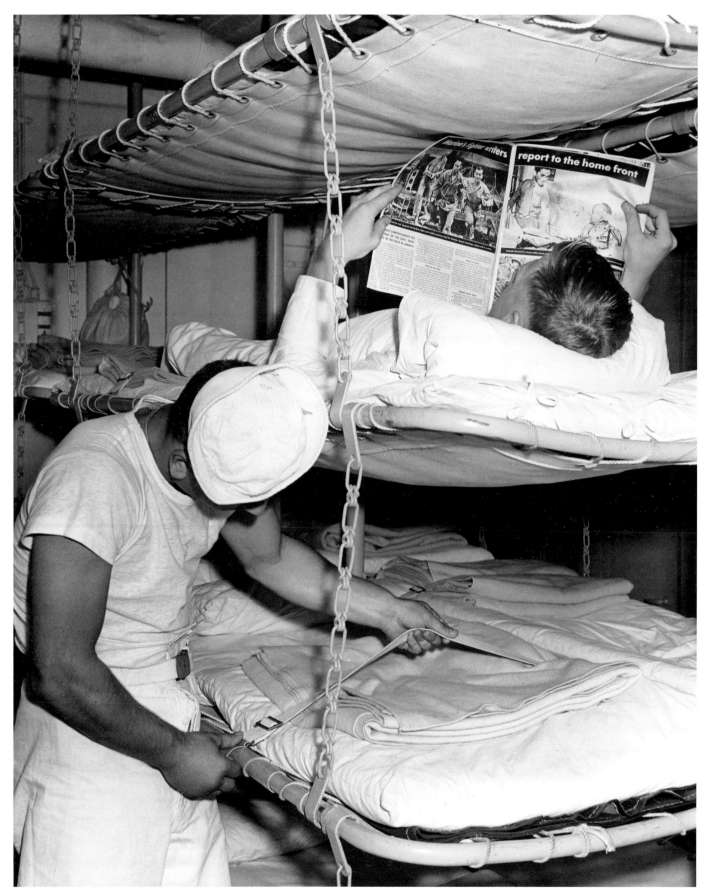
Seamen's berthing in *Missouri*.
NARA

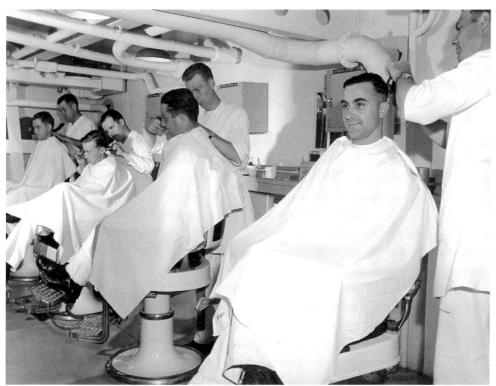

Regulation haircuts being dispensed in the barbershop.
NARA

Shoe cobblers at work.
NARA

Stranding on the Thimble Shoal

On November 1, 1948, *Missouri* sailed from Norfolk for three weeks of exercises in the Arctic with naval cadets embarked. This voyage took her across the Davis Strait separating Canada and Greenland before making another call at Portsmouth and returning for three months of refitting at Norfolk. By now *Missouri* was the only member of her class still in service, with President Truman reluctant to decommission her despite the advice of Secretary of Defense Louis Johnson, Secretary of the Navy John Sullivan, and Chief of Naval Operations Louis E. Denfeld, a decision that no doubt owed much to Truman's personal attachment to the ship. A second training cruise was carried out in August and September 1949, this time to the Mediterranean, before the ship returned to Norfolk for the holiday season.

On January 17, 1950, *Missouri* was scheduled to leave Norfolk to carry out exercises off Guantánamo Bay as usual. Before sailing for Cuban waters, Captain Brown received a request from the Naval Ordnance Laboratory (NOL) to sail down a channel along which acoustic cables had been laid to record the sound of the ship's propellers, part of an effort to detect and identify warships by their characteristic noises. Brown paid little attention to this message and instead focused on navigating the ship to the Windward Passage. But Brown was informed by the ship's navigator, Lt. Cdr. Frank G. Morris, that the test would be carried out on leaving Hampton Roads, just after passing Fort Wool and Old Point Comfort. The entrance and exit of the test channel would be marked by orange and white buoys, between which a series of five other buoys marked the navigation channel. Unfortunately, three beacons had been removed by the time *Missouri* sailed at 0725 on the 17th without this being reported on the ship's information board. At this time *Missouri* had a full supply of ammunition and food and her oil tanks were 95 percent full, giving her a draft of 36 feet forward and 36 feet 9 inches aft. Initially the ship steamed under pilot, but at 0749 he was landed after passing buoy no. 3 on the Elizabeth River and control of the vessel passed to Brown, who went on to 10 knots and steered 053 degrees in clear weather and good visibility.

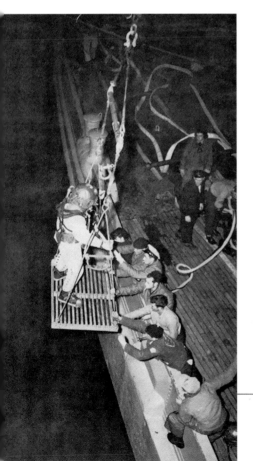

A diver prepares to examine the hull of *Missouri* as members of her crew look on.
USN

Missouri stranded on the Thimble Shoals in January 1950.
NARA

A diver prepares to examine the hull of *Missouri*.
USN

At 0805 *Missouri* passed Fort Wool, after which Brown went to the chart house and announced that the ship would soon enter the acoustic channel. After two minutes Brown appeared on the navigating bridge and informed the two lieutenants on watch that the trial was about to begin. At 0810, with the ship on a course of 060 degrees, Brown made his way to 08 Level bridge in the forward superstructure. *Missouri* was then taken by a strong current and the course altered to 058 degrees, at which point several officers began expressing their concern as to the proximity of the Thimble Shoal. Nonetheless, against Morris' advice, Brown increased speed to 15 knots, at which point the first buoy was sighted at a distance of one thousand yards, Brown steered to pass it to port when he should have passed it to starboard. Those on watch quickly pointed out other buoys Brown was certain marked the test channel. This was not the case since they actually marked a fishing bank in shallow water. Recognizing the error, several officers pointed this out to Brown, who only reduced speed. Meanwhile, those operating the navigation radar had also realized the ship was entering hazardous water but said nothing, believing their equipment to be defective. Aside from this the ship's sonar was out of order. At 0815 the navigator became aware of the danger and issued a warning to Brown, who was convinced Morris had miscalculated the position of the ship. Brown, though, only ordered the helmsman to bring the ship slightly to starboard. At 0817, *Missouri* took ground violently on a mudflat 1.6 miles from the Thimble Shoal Light and during an exceptionally high tide. The hull was raised several feet above the waterline and sand had entered the condenser intakes, requiring the engines to be stopped. At 0830, the Norfolk Naval Base was informed of the incident and all available tugs were ordered to Fort Monroe. *Missouri* was found to have run over eight hundred yards across the Thimble Shoal and it would be no simple matter to refloat the 57,000-ton giant.

U.S. Navy officials wanted to use private companies to carry out the work, but this was strongly opposed by Rear Adm. Allan E. Smith, Atlantic forces commander, who had been warned that the progress of his career depended on the outcome of the salvage operation. This was particularly the case since the domestic and international

Rear Adm. Homer N. Wallin who led the refloating operations. USN

Stores being unloaded using a floating crane. USN

media had got hold of the story, resulting in 10,000 letters from civilians containing suggestions on how to refloat the ship. An article appeared in the Russian media proclaiming "*Missouri*, a failed demonstration of the low level of American naval technology. The level of education and qualifications of the crews are far from exemplary." Nor did the U.S. press lose the opportunity for good copy: "Join the Navy and See the Thimble Shoal."

The officer charged with refloating *Missouri* was Rear Adm. Homer N. Wallin, who had supervised salvage operations after the attack on Pearl Harbor. His plan of action was to lighten the vessel as much as possible, level the surrounding area, excavate a clearance channel, and finally gather a large number of tugs to drag the vessel off the mudflat.

Ammunition being transferred into specially prepared barges.
USN

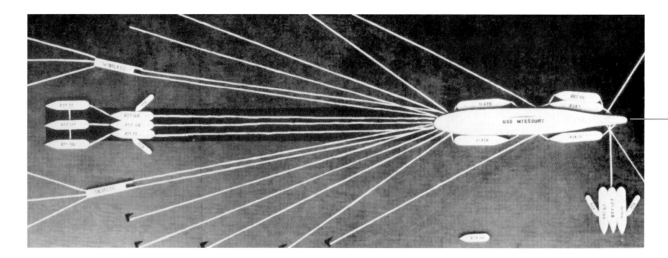

Since the vessel had grounded at high tide with a high coefficient, a date with the same conditions had to be chosen, and that meant February 2. On January 19, the Army dredger *Comber*, joined three days later by the civilian dredger *Washington*, set about creating a clearance channel forty feet deep and one hundred feet wide on either side of the vessel. Meanwhile, the ammunition, an anchor and its chain, and eight thousand tons of fuel had also to be unloaded, the oilers *Chemung* (AO 30) and *Pawcatuck* (AO 108) being mobilized to transfer the oil as ships were moored nearby to deliver the power necessary to operate the ship. Divers were also needed to create openings in the mud to eliminate the suction effect when the towing started. Seventy-five-pound charges of explosives were also strategically inserted in the most densely packed mud to facilitate dredging.

A first effort at towing on the 31st ended in failure, with the ship's bottom being punctured by an old anchor. At dawn on February 1, though, after several minutes of sustained effort, a concerted pull by five tugs moored alongside the ship, five pulling to starboard, six towing the stern, and seven responsible for keeping the main tugs in alignment, assisted by ten anchors, succeeded in bringing *Missouri* off the Thimble Shoal at 0709. *Missouri* duly hoisted her ensign, and sirens blared as she was brought back into Norfolk. It remained to examine the damage in one of the basins of the yard, the most suitable being occupied by the unfinished hull of the *Kentucky* (BB 66), which was quickly removed to allow her sister to undergo repairs.

The complex arrangements for hauling *Missouri* off the shoal.
USNHC

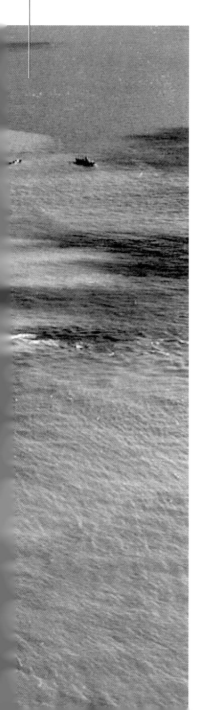

Missouri goes into dry dock at Norfolk for repairs.
USN

The damage was found to be relatively minor. The starboard bulge had been driven in at frames 99 and 114 and some fuel bunkers required refitting. The bilge keel had suffered some damage and the condenser heads had to be opened and cleaned. In addition, the hull and three of the propellers had suffered slight damage from the towing operations, as had been the case with some of the fairleads, railings, and other deck fittings, while several of the engine room gauges had been shattered by the explosive charges.

Captain Brown and four officers, including Morris, were court-martialed. Brown eventually accepted sole responsibility and was relieved of his command, subsequently attempting suicide while his wife suffered a nervous breakdown. He spent the rest of his career on shore duty. Two officers were acquitted, one received a letter of reprimand, and another was demoted in the seniority list. These, however, were not the only consequences. Persistent though baseless rumors followed *Missouri* for thirty years to the effect that the faceplates of turret 3 had been dislodged and the ship's speed was limited to 15 knots, claims that may have hindered her being selected for reactivation in later years.

Meanwhile, in an effort to silence her critics, *Missouri* successfully carried out five days of sea trials from February 8 under the temporary command of Brown's predecessor, Capt. Harold P. Smith, who was able to report that all damage sustained during the stranding and salvage operations had been repaired by the Norfolk Naval Shipyard.

Secretary of Defense Louis A. Johnson then ordered *Missouri* to become a training ship for midshipmen and reservists. With a crew of 1,500, she participated in the Portex and Caribex 50 exercises held in March, and beginning on June 25 she carried out her first cruise of the Caribbean with Navy midshipmen embarked before putting in at Halifax and Boston. On August 13, Smith's successor, Capt. Irving T. Duke, was informed his ship was combat ready.

Extensive damage to *Missouri*'s hull caused by the stranding.
USN

Rear Adm. Homer N. Wallin points out some of the damage.
USN

Korea

The dramatic events unfolding in Korea had revealed the need for significant fire support for U.S. ground troops. This focused attention on the *Iowa*-class battleships with their ability to deploy their formidable guns without hindrance, and *Missouri* was the first unit of the class to reach Southeast Asia.

After five days of replenishment at Norfolk, *Missouri* sailed for Limon Bay at the eastern entrance to the Panama Canal on August 18, 1950. Proceeding at 25 knots, she found herself in a storm that worsened into a hurricane and caused significant damage to the ship, wrecking six of her boats, putting a 40-mm mount out of service, and washing two HO3S-1 helicopters overboard.

After traversing the Panama Canal on the 23rd and refueling at Balboa, *Missouri* reached Pearl Harbor eight days later, where repairs were carried out to her hurricane damage. On September 5 she weighed anchor and headed for the Japanese port of Sasebo but had to alter course several times to avoid Typhoon Kezia. On the 14th she joined the destroyers *Samuel N. Moore* (DD 747) and *Maddox* (DD 731), with Rear Adm. Allan E. Smith commanding Task Force 95 shifting his flag from the latter to *Missouri*. That same day she fired fifty-two 16-inch shells on bridge and railway targets near Samchok. Joined by *Helena* (CA 75) and the destroyer *Brush* (DD 745), in the early hours of the 16th *Missouri* bombarded Pohang in support of troops of the 3rd Division of the army of the Republic of Korea (ROK), hard-pressed on the banks of a river. With enemy positions pulverized by 298 heavy-caliber shells, the ROK troops were able to extricate themselves and cross a bridge to safety. At around 1000 the destroyer tender *Dixie* (AD 14) embarked Rear Admiral Smith as *Missouri* headed for Sasebo for replenishment. By the 20th she was back on the gun line in support of the U.S. 7th Infantry Division, which had landed at Inchon as part of Operation Chromite. The following afternoon General MacArthur came on board *Missouri* accompanied by the commander in chief of the Seventh Fleet, Vice Adm. Arthur D. Struble, and Maj. Gen. Edward M. Almond of the U.S. Army X Corps to confer on the situation at the front.

Missouri refueling the destroyer *Joseph P. Kennedy Jr.* (DD 850) in August 1950.
USN

Replenishment at sea.
NARA

Missouri anchored at Inchon among a large concourse of ships of all types on September 15, 1950.
USN

After escorting the carrier *Valley Forge* (CV 45) off the east coast of Korea, in mid-October *Missouri* carried out a major bombardment of the Mitsubishi factories at Chongjin in which the cruisers *Helena* (CA 75), *Worcester* (CL 144), and HMS *Ceylon* and a number of Canadian, Australian, and British destroyers also participated. With this *Missouri* put in at the port of Wonsan for replenishment, during which the indefatigable Bob Hope came on board to entertain the crew. By this time *Missouri* had destroyed fifteen bridges, two tunnels, thirty-three railroad lines, and numerous artillery and command positions.

On October 19, however, some 380,000 Chinese soldiers under General Peng Dehuai launched a major counteroffensive against UN troops, who were forced to fall back on Hungnam. *Missouri* covered the evacuation between December 22 and 24 and fired 162 16-inch and 699 5-inch shells. In late January *Missouri*, together with the light cruiser *Manchester* (CL 83), participated in Operation Ascendant near Kansong-Kosong, during which she fired 382 16-inch and 3,002 5-inch shells. After putting in at Sasebo for replenishment, by February 5, 1951, she was back in support of the U.S. Army 10th Corps at Kangnung before withdrawing to Pusan on the 23rd, where Syngman Rhee, U.S. ambassador to Korea John J. Muccio, Korean navy chief of staff Rear Admiral Sohn Won-yil, and other important figures were received on board.

On March 11, *Missouri* bombarded railroad installations and enemy positions at Chaho, Chongjin, and Wonsan, and on the 24th she sailed for Yokohama after being relieved by *New Jersey*. Her deployment over, on the 28th she began the long voyage home to Norfolk, which she reached on April 27 via Long Beach and the Panama Canal.

Missouri and *New Jersey* at Yokohama.
NARA

Missouri bombarding Kansong.
USNHC

During her first deployment in Korea *Missouri* had fired 2,895 16-inch and 8,043 5-inch shells and had steamed 62,100 miles.

Rear Adm. James L. Holloway, commander in chief of the Atlantic Fleet, embarked in *Missouri* on her return to Norfolk, and numerous training cruises were carried out that summer and fall in the North Atlantic with stops at Cherbourg, Oslo, New York, and Colón at the entrance to the Panama Canal. The ship was docked at Norfolk between October 18, 1951, and January 30, 1952, and the following day sailed for six weeks of exercises in Guantánamo Bay.

After several uneventful sorties, *Missouri* returned to Norfolk Naval Shipyard to prepare for her second deployment to Korea. On September 11 *Missouri* sailed from Hampton Roads for the combat zone, reaching Yokosuka on October 17, where two days later she became the flagship of the Seventh Fleet.

Missouri at Long Beach in 1951.
NARA

Missouri on arrival at Norfolk on April 27, 1951.
USN

Missouri on arrival at Norfolk on April 27, 1951.
USN

Iowa and *Missouri* in October 1952.
NARA

Missouri opens fire on communist positions. NARA

Capt. Warner Edsall, who succumbed to a heart attack while maneuvering his ship off Sasebo. USNHC

On November 17, *Missouri* participated in the defense of Chongjin and on December 8 bombarded areas in the vicinity of Tanchon and Songjin. Logistical support occupied most of *Missouri*'s time, shelling the towns of Chaho, Wonsan, Hŭngnam, and Kojo. On December 21, one of her two helicopters came down at sea off Hŭngnam with the loss of her three crewmen. On January 5, 1953, *Missouri* put in at Sasebo where she received a visit from Gen. Mark Clark, commander in chief of UN forces, and Admiral Sir Guy Russell, commander in chief of the British Far East Fleet. On the 23rd she again shelled Wonsan, Tanchon, Hŭngnam, and Kojo before carrying out a succession of patrols and bombardment operations along the east coast of Korea, as had been the case in 1950–51.

Another tragedy occurred on March 5, when Capt. Warner Edsall succumbed to a heart attack while maneuvering his ship at Sasebo. On the 25th *Missouri* fired her last shells against Kojo and was relieved by *New Jersey* on the 6th, reaching Norfolk on May 4 to a rapturous welcome. Resuming her training duties, on June 8 she put to sea on a monthlong cruise with stops at Trinidad, Cuba, and Panama. *Missouri* then underwent a lengthy refit at Norfolk between November 20, 1953, and April 4, 1954. On June 7 she began what was destined to be her last training cruise with stops at Lisbon and Cherbourg, but before departing she and her three sisters carried out exercises off the Virginia Capes, the only occasion on which they sailed in company. Earmarked for deactivation, *Missouri* sailed from Norfolk on August 23 and put in at Long Beach and San Francisco before reaching Seattle on September 15. Three days later she entered Puget Sound Naval Shipyard for mothballing.

A five-month inactivation process ended with a ceremony on February 26, 1955, by which she joined the Pacific Reserve Fleet, the beginning of a thirty-year hibernation.

Missouri in 1954.
USN

Missouri berthing at Norfolk with the assistance of tugs in August 1954. The carrier *Saipan* is to starboard of her, and *Iowa* is moored on the other side of the pier.
USN

Missouri decommissioned at Bremerton in 1963.
NARA

Civilian visitors on *Missouri's* forecastle in September 1975.
USN

Missouri at the Service of Presidents Reagan and Bush

By the 1980s practically every battleship had disappeared from the lists of the world's combat fleets, the death knell having been sounded decades earlier by the aircraft carrier and the submarine. Few imagined they would ever see the super-dreadnoughts of an earlier era in open water again, but what saved the sole survivors of the *Iowa* class from certain demolition was their formidable 16-inch armament.

As were her sisters, *Missouri* was destined to be refitted in the early 1980s. Despite her advanced age, *Missouri* had only been on active service for ten years and one month, whereas the class was designed with a service life of thirty-five years.

On May 14, 1985, she left Bremerton under tow by the salvage and rescue ship *Beaufort* (ATS 2), arriving at Long Beach Naval Shipyard on the 25th.

During the following months all antiaircraft weapons were landed together with much obsolete electronic equipment. Armored box launchers (ABLs) housing the Tomahawk cruise missiles were positioned between and abreast the two stacks facing aft. Sixteen Harpoon missile launchers were mounted on either side of the after stack together with high-performance 20-mm Phalanx CIWS guns in the after and forward superstructure. The ship was also fitted with the latest radar as well as new detection systems. Provision was made for operating drones from the fantail, and it was decided to restore the ship's 794-pound bell, which had been entrusted to Jefferson City, Missouri, to celebrate the 150th anniversary of the state in 1971.

On January 29, 1986, *Missouri* carried out her first trials off San Francisco and was commissioned on May 10, 1986, under Capt. Lee Kaiss, who holds the distinction of being the only officer in the history of the U.S. Navy to command the ship at the time of her commissioning, or recommissioning in this case, and her final deactivation on March 31, 1992. During the ceremony Defense Secretary Caspar W. Weinberger spoke these words to the ten thousand people gathered: "This is a day to celebrate the rebirth of American sea power. Listen for the footsteps of those who have gone before you. They speak to you of honor and the importance of duty. They remind you of your own traditions." Also present that day was the Governor John Ashcroft of Missouri; U.S. Senator Pete Wilson of California; Secretary of the Navy John Lehman; Mayor Dianne Feinstein of San Francisco; and the ship's original sponsor in 1944, Margaret Truman, who issued this injunction to Capt. Kaiss and his men: "Please take good care of my baby."

Missouri being readied to enter Long Beach Naval Shipyard for modernization on May 25, 1984.
NARA

Missouri in dry dock during her refit at Long Beach.
USN

Missouri's recommissioning ceremony on May 10, 1986.
USN

Between May 14 and 25, the ship completed various tests for the Board of Inspection and Survey off Southern California, and in early August was again at sea for crew training purposes. On September 10 *Missouri* sailed from Long Beach on a three-month endurance cruise involving a circumnavigation of the world similar to that carried out eighty years earlier by her namesake, the pre-dreadnought battleship *Missouri* (BB 11), as part of the Great White Fleet. The voyage took her to Pearl Harbor, Sydney, where she participated in the seventy-fifth anniversary of the Royal Australian Navy, Diego Garcia in the Indian Ocean, through the Suez Canal and on to Istanbul, Lisbon, and back to Long Beach on December 19 after traversing the Panama Canal.

After a spell in dockyard hands, in June 1987 *Missouri* began preparing for a six-month mission in the Indian Ocean before contributing to Operation Earnest Will, the escort of tanker convoys in the Persian Gulf. She sailed on July 25 and arrived on station on August 30, where she joined Task Group 70.10 that included the carrier *Ranger* (CV 61), the guided-missile cruisers *Long Beach* (CGN 9), *Richmond K. Turner* (CG 20), *Gridley* (CG 21), and *Bunker Hill* (CG 52), the destroyers *Hoel* (DDG 13), *Buchanan* (DDG 14), *John Young* (DD 973), and *Leftwich* (DD 984), the frigates *Curts* (FFG 38) and *Robert E. Peary* (FF 1073), the ammunition ship *Shasta* (AE 33), and the fleet oilers *Assayampa* (AO 145), *Wichita* (AOR 1), and *Kansas City* (AOR 3).

The protection of tankers sailing from Kuwait kept *Missouri* at sea for more than a hundred days, during which she was threatened by Iranian mines and powerboats requiring all combat systems to be kept on high alert beyond the Strait of Hormuz. It was therefore with a sense of achievement that *Missouri* turned for California on December 2, reaching Long Beach on January 19, 1988, after calling at Diego Garcia and Pearl Harbor.

After refitting at Long Beach, in June and July *Missouri* took part in the RimPac 88 exercise off Hawaii with vessels of the Japanese, Canadian, and Australian navies and involving more than 50,000 men. During these maneuvers *Missouri* was usually escorted by a *Ticonderoga*-class guided-missile cruiser, a *Kidd*-, *Arleigh Burke*-, or *Spruance*-class guided-missile destroyer, three *Oliver Hazard Perry*–class guided-missile frigates and an oiler. *Missouri* spent the following summer tied up at Pier D in Long Beach, but an otherwise quiet interlude was disrupted in July by the filming of Cher's sensational video *If I Could Turn Back Time*, the singer cavorting around turrets 1 and 2 in an exiguous outfit. This began something of a trend, because in 1992 *Missouri* was the centerpiece of Andrew Davies' movie *Under Siege* with Steven Seagal and Erika Eleniak. In 2001, with *Missouri* by now in a state of preservation, several scenes of Michael Bay's *Pearl Harbor*, starring Ben Affleck and Kate Beckinsale, were shot on board. Her latest appearance is in Peter Berg's *Battleship*, with Taylor Kitsch and Rihanna.

In September and October 1991, *Missouri* was in the Pacific with *New Jersey* and fifty-four U.S., Canadian, Korean, and Japanese warships to participate in PACex 89.

Missouri passing under the Golden Gate Bridge on May 10, 1986.
USN

The oiler *Kawishiwi* refueling *Missouri* and the carrier *Kitty Hawk* in July 1986.
USN

Missouri in Sydney in October 1986.
NARA

Two of the armored box launchers await their Tomahawk cruise missiles.
USN

Missouri during a gunnery exercise in 1988.
USN

Missouri drydocked at Long Beach in 1990.
USN

Missouri drydocked at Long Beach in 1990.
USN

Missouri participating in the RimPac 90 exercise off Hawaii.
USN

Further exercises followed in early 1990, and on June 13 Captain Kaiss was again given command of the ship. *Missouri* participated in the RimPac 90 exercise and it was intended for her to make several stops at ports along the West Coast of the United States during September, but the invasion of Kuwait by Iraqi president Saddam Hussein on August 2 scuttled these plans. The smaller Kuwaiti army was quickly overwhelmed and the occupation was completed in four hours. This aggression was quickly condemned by the United Nations Security Council, and on August 6 President George H. W. Bush approved the commencement of Operation Desert Shield.

So it was that *Missouri* again found herself involved in a new conflict, and after refueling at Long Beach she sailed for the Persian Gulf in a blaze of publicity on November 13. Steaming westward via Hawaii, the Philippines, and Thailand she passed the Strait of Hormuz on January 3, 1991, and joined the naval forces mustered for the coming Operation Desert Storm, including 115 U.S. Navy and 50 coalition ships. At 0140 on January 17 *Missouri* launched her first Tomahawk cruise missiles at enemy positions, and by the 22nd she had fired twenty-seven against Iraqi fortified and strategic positions.

On the 29th, the guided-missile frigate *Curts* (FFG 38) cleared the way for *Missouri* to move into the Northern Persian Gulf, and from February 3 she fired 112 16-inch shells against fortified and command positions near the Saudi border. She was relieved by *Wisconsin* two days later but bombardment missions continued, including the shelling of the border town of Ras Al Khafji on the night of 11th and 12th during which 60 shells were fired. On the 23rd she supported an amphibious raid with 133 16-inch shells and the following day silenced artillery batteries on the island of Faylaka. No sooner had the shelling stopped than the ship's drone spotted hundreds of Iraqi soldiers waving white flags, unquestionably the first case of surrender to an unmanned aircraft.

Needless to say, *Missouri* represented a high-profile target for enemy missile batteries, and on the 25th two HY-2 Silkworm antiship missiles were fired against her while she was under escort by the guided-missile frigate *Jarrett* (FFG 33) and the British guided-missile destroyer HMS *Gloucester* (D 96). The first missile failed to reach *Missouri* but the second had to be destroyed by a Sea Dart surface-to-air missile launched by the *Gloucester*, the missile exploding on contact with the sea seven hundred yards beyond its intended target. During this attack the firing of the *Jarrett*'s Phalanx CIWS resulted in a number of projectiles striking *Missouri*, with minor damage caused to the superstructure and shrapnel injury to a sailor. *Jarrett* was 3,500 yards from *Missouri* at the time and an inquiry failed to prove whether a Phalanx operator had fired accidentally.

Missouri completed her combat mission on March 21. Aside from the missile attack on February 25, the greatest threats to the ship during Operation Desert Storm were mines, about fifteen of which were destroyed by the ship's 5-inch or Phalanx guns or Marine rifles. The ship fired a total of 783 16-inch shells and 28 Tomahawks during operations.

Missouri returned home via Fremantle, Hobart, and Pearl Harbor, reaching Long Beach on May 13.

A mine spotted near *Missouri* during Operation Desert Shield in 1991.
USN

Missouri opens fire on Iraqi positions on February 6, 1991.
USN

Missouri in the Persian Gulf in February 1991.
USN

Wisconsin and *Missouri* (furthest from the camera) undergoing replenishment from the fast combat support ship *Sacramento* (AOE 1), whose engine plant had originally been installed in their unfinished sister *Kentucky*.
USN

The launching of a Tomahawk during Operation Desert Storm.
NARA

The main battery shelling Iraqi positions on February 6, 1991.
USN

Missouri in 1991.
J. PRADIGNAC

After an inevitable spell of refitting, the time came for *Missouri* to carry out her last important cruise. On November 29, 1991, *Missouri*, the battleship upon which the Japanese Instrument of Surrender had been signed in 1945, sailed from Long Beach for Pearl Harbor to commemorate the fiftieth anniversary of the attack that brought the United States into World War II. Although she departed without fanfare, neither the crew nor those ashore had any doubt as to the significance of the moment. "It's sad," said Lt. Wes Carey, one of the ship's 1,500 crewmen. "It's the end of an era."

Moored at Pier 8 at Pearl Harbor dockyard, *Missouri* overlooked a platform on which President George H. W. Bush gave a speech to a large crowd that included two thousand veterans. The former Navy pilot, who had been shot down by Japanese antiaircraft fire in the Pacific in September 1944, uttered these words: "Well, let me tell you how I feel. I have no rancor in my heart towards Germany or Japan, none at all. And I hope, in spite of the loss, that you have none in yours. This is no time for recrimination." The president was then welcomed on board *Missouri* and had a moment to reflect before the famous plaque commemorating the surrender ceremony on September 2, 1945.

After *Missouri*'s return to Long Beach on December 20, her men could begin the long process of retiring her from service.

Missouri entering Pearl Harbor to commemorate the fiftieth anniversary of the Japanese attack. Note the USS *Arizona* Memorial in the background.
USN

Missouri approaching Pearl Harbor under tow for preservation as a museum close to the USS *Arizona* Memorial. USN

Arizona's Neighbor

The budget cuts of the early 1990s resulted in *Missouri* being definitively deactivated at Long Beach on March 31, 1992. In the course of a moving ceremony, Capt. Lee Kaiss addressed the crew as follows:

> Our final day has arrived. Today the final chapter in battleship *Missouri*'s history will be written. It's often said that the crew makes the command. There is no truer statement [...] for it's the crew of this great ship that made this a great command. You are a special breed of sailors and Marines and I am proud to have served with each and every one of you. To you who have made the painful journey of putting this great lady to sleep, I thank you. For you have had the toughest job. To put away a ship that has become as much a part of you as you are to her is a sad ending to a great tour. But take solace in this—you have lived up to the history of the ship and those who sailed her before us. We took her to war, performed magnificently and added another chapter in her history, standing side by side with our forerunners in true naval tradition. God bless you all.

In doing so he became the last captain of the last battleship of the U.S. Navy, and the last operational vessel of her type in the world. *Missouri* was then placed in the Reserve Fleet at Puget Sound Naval Shipyard, Bremerton, Washington. She remained on standby until January 12, 1995, before being stricken from the Navy List. The only hope for her survival was to become a floating museum, the first of the *Iowa* class to be preserved.

On May 4, 1998, Secretary of the Navy John H. Dalton signed the agreement donating the battleship to the *Missouri* Memorial Association of Honolulu, Hawaii. The USS *Missouri* Memorial Association, Inc., is a nonprofit organization whose mission is to operate and maintain a national monument commemorating the end of World War II and to pay lasting tribute to the role of the U.S. Navy in establishing world peace. The *Missouri* Memorial Association is exclusively for charitable, scientific, literary, or educational purposes and receives no funding whatsoever from the U.S. Navy or any other government agency for the operation and ongoing maintenance of the vessel.

On May 23, *Missouri* was towed from Bremerton to Astoria, Oregon, where she was docked before being towed across the Pacific to her final resting place 550 yards from the wreck of the *Arizona*. The National Park Service, which is responsible for the Battleship *Arizona* Memorial, was concerned that the presence of *Missouri*, which symbolizes the victory of the United States over the Japanese Empire, would overshadow the *Arizona*, whose destruction and sinking on December 7, 1941, represents the sacrifice of many sailors on that day of infamy. It was therefore decided to position *Missouri* with her bow closest to the remains of the *Arizona* so that the ceremonies regularly held on the fantail would not disturb the meditation of those paying their respects over the wreck. The *Missouri* Memorial opened to the public on January 29, 1999, and on October 23, 2003, was honored with a visit by President George W. and First Lady Laura Bush.

On October 19, 2009, *Missouri* was docked for the installation of a new anticorrosion system and for maintenance work on the vessel as a whole. The total cost of the operation was $18 million, following which *Missouri* was reopened to the public from January 30, 2010.

Missouri entering Pearl Harbor under tow on June 21, 1998.

The Career of the Battleship USS *Missouri* (BB 63) 455

Missouri approaching the wreck of the *Arizona*.
USN

The carrier *Nimitz* passing *Missouri* on October 27, 2003.
USN

Battleship *Missouri* Memorial
63 Cowpens St.
Honolulu, HI 96818
www.ussmissouri.org

USS *MISSOURI* (BB 63)

Commissioned	June 11, 1944
Decommissioned	February 26, 1955
Commissioned	May 10, 1986
Decommissioned	March 31, 1992

Missouri drydocked at the Pearl Harbor Naval Shipyard for a three-month refit costing a total of $18 million.
USN

Missouri being towed to her final berth off Ford Island.
USN

SIXTEEN

The Career of the Battleship USS *Wisconsin* (BB 64)

The "Wisky"

The keel of *Wisconsin* (BB 64) was laid at Philadelphia Naval Shipyard on January 25, 1941. The launch ceremony was held on December 7, 1943, under the direction of Rear Adm. Milo F. Draemel, commander of the yard. Also present were the Secretary of the Navy Ralph A. Bard and numerous naval officers, with the ship being sponsored by the wife of the governor of the State of Wisconsin, Mrs. Walter S. Goodland. The ways had to be lubricated with more than forty-five tons of grease for the ship to take to the water.

Although two lives had been lost in the construction of the hull, the fitting-out process was completed uneventfully at Pier 4 and the ship was commissioned on April 16, 1944, after thirty-nine months of labor. Before a 1,500-strong crowd, Robert M. La Follette Jr., and representatives of Congress, Old Glory was hoisted at the stern of the vessel by Admiral Draemel at 1408 precisely. The industrialist Carl Pick of West Bend, Wisconsin, offered the battleship a commemorative plaque manufactured in his foundries from her earlier namesake, the predreadnought *Wisconsin* (BB 9). The event was attended by ten veterans of BB 9 including the new battleship's first commander, Capt. Earl E. Stone, who had sailed in her as a cadet in 1916.

Wisconsin under construction on January 12, 1943.
USN

The boilers in place on July 8, 1943.
USN

Wisconsin shortly before launching.
USN

Wisconsin on the slip on her launch day at the Philadelphia Naval Shipyard, December 7, 1943.
USN

The First Lady of Wisconsin, Mrs. Walter S. Goodland, christens the ship before launch.
USN

Wisconsin slips down the ways on December 7, 1943.
NARA

Tugs prepare to tow *Wisconsin* to the fitting-out pier shortly after launch. USN

Wisconsin fitting out in 1944.
USN

The crew mustered for inspection by the captain.
USN

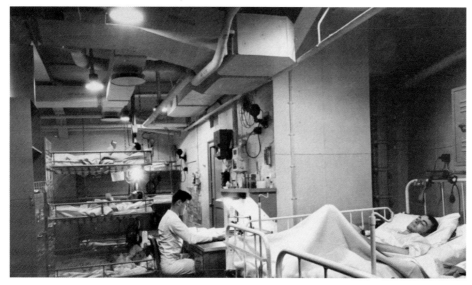

The ship's sickbay.
USN

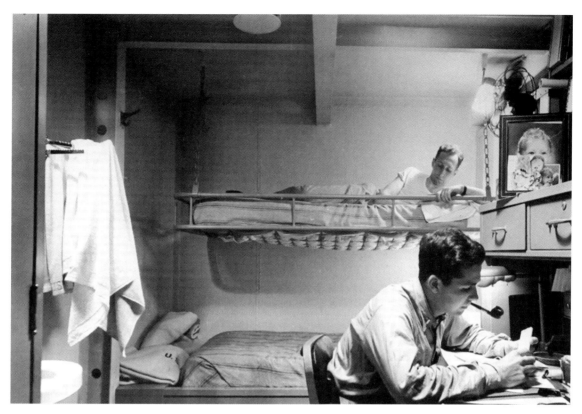

Officers' quarters in *Wisconsin*.
USN

After provisioning and several brief sallies in Chesapeake Bay, *Wisconsin*, by now affectionately known as "Wisky," sailed for Norfolk on July 7 to carry out her steam trials in the Caribbean, mainly off Trinidad and in the Gulf of Paria, the ship occasionally returning to Philadelphia for technical reasons. These included tests of the floatplane catapults, towing drills, gunnery exercises, and full-power trials together with a wide range of maneuvers typical of this intense training period. During these trials the officers noted that *Wisconsin* could make 27.5 knots with only four boilers lit.

On September 24 *Wisconsin* sailed for the East Coast, traversing the Panama Canal for the first time during which she had just a few inches of clearance on either side. Numerous fenders and other measures were installed along the ship's side but some 120 feet of paint had been scraped off by the time she exited the lock.

Now up to full war complement, the ship felt cramped after the addition of antiaircraft armament had increased the number of those on board from a design complement of 2,000 to 2,724 by the end of 1944. With every conceivable space occupied and no air conditioning fitted, it was not unusual to find men sleeping out on deck.

A compensating factor was the quality of the offerings of the ship's galleys, where some 4,100 pounds of vegetables, 1,640 pounds of fruits, 2,464 pounds of meat, 163 pounds of butter, 1,500 pounds of flour, 540 dozen eggs, and 200 pounds of coffee were prepared each day. Morale was good and the ship had an excellent boxing team.

On October 2 she was assigned to Adm. William F. Halsey's Third Fleet in the Pacific, reaching the anchorage of Ulithi in the Caroline Islands on December 9 and joining Task Force 38, commanded by Vice Adm. John S. McCain and composed of the carriers *Essex* (CV 9), *Yorktown* (CV 10), *Ticonderoga* (CV 14), *Lexington* (CV 16), *Wasp* (CV 18), and *Hancock* (CV 19), the escort carriers *Independence* (CVL 22), *Cowpens* (CVL 25), *Monterey* (CVL 26), *Langley* (CVL 27), *Cabot* (CVL 28), and *San Jacinto* (CVL 30), the battleships *North Carolina* (BB 55), *Washington* (BB 56), *South Dakota* (BB 57), *Massachusetts* (BB 59), and *Alabama* (BB 60), the heavy cruisers *Louisville* (CA 28), *New Orleans* (CA 32), and *Baltimore* (CA 68), the light cruisers *San Diego* (CL 53), *Santa Fe* (CL 60), *Mobile* (CL 63), *Vincennes* (CL 64), *Pasadena* (CL 65), *Biloxi* (CL 80), *Miami* (CL 89), *Astoria* (CL 90), and *Oakland* (CL 95), and twenty-eight destroyers.

Wisconsin shortly after commissioning.
USN

Wisconsin lying alongside the salvaged battleship *Oklahoma* at Pearl Harbor in November 1944.
USN

Wisconsin at the time of her steam trials.
USN

On December 11, 1944, TF 38 sailed from Ulithi to carry out a raid against Luzon (Operation Love III). After refueling 350 miles east of its target, the air wing launched attacks against Clark Field, Angeles Field, Masinloc, Manila Bay, San Fernando, and Cabatuan during which the Japanese lost 27 ships and 269 aircraft. *Wisconsin*'s first sortie with the fleet, however, was made memorable by a typhoon in mid-December during which three destroyers were sunk and many ships damaged. *Wisconsin* had one of her Kingfisher floatplanes ripped off its catapult and it experienced valve sealing problems in boilers 3 and 4 in fireroom 2. The necessary repairs required shaft 2 to be immobilized for two hours.

After refueling 250 miles east of the Philippines on the 19th, TF 38 gratefully reached Ulithi five days later. But it was back at sea on the 30th to attack Japanese air bases on Formosa on January 3 and 4, 1945. On the 6th the fleet took up a position southeast of Cape Engano and put in fresh attacks against Luzon, Formosa, and the Pescadores Islands. The 12th saw the beginning of Operation Gratitude, the bombardment of Japanese positions in French Indochina, with further bombardments of Formosa and attacks on Hong Kong, Canton, Hainan, Sakishima, Okinawa, and Ryūkyū followed by reconnaissance missions over Chimu Bay and Nakagusuku Bay. On the 20th, *Wisconsin*'s floatplanes sortied to rescue a pilot who had ditched, but nothing was found. The following day her hospital received seventeen casualties from the destroyer *Maddox* (DD 731) after she was struck by a kamikaze off Formosa, a change from the usual routine of appendectomies and shipboard accidents. By the 25th the fleet was back in Micronesia, *Wisconsin* having sailed 25,706 miles in fifty-eight days at sea.

On January 26 *Wisconsin* was assigned to Task Force 58, commanded by Vice Adm. Marc A. Mitscher in *Missouri*. She formed part of Rear Adm. Ralph E. Davison's Task Group 58.2, composed of the carriers *Lexington* (CV 16) and *Hancock* (CV 19), the escort carrier *San Jacinto* (CVL 30), the heavy cruisers *Boston* (CA 69) and *Pittsburgh* (CA 72), together with an escort of thirteen destroyers. On February 10, *Wisconsin*, now flying the flag of Rear Adm. Edward W. Hanson commanding the 9th Battleship Division, weighed anchor with her Task Group and sailed for the coast of Japan. On the 13th, TG 58.2 refueled at sea four hundred miles northeast of Luzon, after which a series of air raids code-named Operation Jamboree were launched against Tokyo on the 16th and 17th. That same February 17 a bombardment of Iwo Jima was carried out in preparation for the landings planned two days later. TG 58.2 remained in support of the landings until the 23rd, cruising at low speed sixty-five miles northwest of Iwo Jima before sailing for a rendezvous two hundred miles east of Iōtō on the 24th. Progress was hindered by a severe storm that obliged TF 58 to reduce speed to 16 knots and left many ships damaged by forty-foot waves, with *Wisconsin* having two 40-mm mounts put out of action. TF 58 then proceeded to a position 190 miles southeast of Tokyo to launch a series of new raids against the city between the 24th and 26th. A final attack against Okinawa on March 1 brought the fleet back to Ulithi on the 4th.

Wisconsin during the great typhoon of December 1944 during which she lost one of her Kingfisher floatplanes.
USN

Wisconsin refueling the destroyer *Halsey Powell* (DD 686). NARA

Still at Ulithi, the crew was watching a film on the evening of the 11th when the anchorage came under attack by three kamikaze aircraft (Operation Tan No. 2), which struck the carrier *Randolph* (CV 14) moored beside *Wisconsin*, resulting in the death of 27 sailors and injuries to 107 others.

Three days later the Fifth Fleet sailed from Ulithi and shaped a course for Ryūkyū. On the 18th and 19th it was the turn of Honshū, Shikoku, Kure, and Kōbe to receive the attentions of U.S. naval aviation, with damage inflicted on the battleships *Yamato*, *Ise*, *Hyuga*, and *Haruna*, the carriers *Amagi*, *Kaiyo*, and *Ryuho*, and many other smaller ships in the Inland Sea. These raids were not without danger given the ferocity of Japanese counterattacks, particularly by kamikaze aircraft, which caused major damage to the carriers *Wasp* (CV 18) and *Franklin* (CV 13). The latter, which lost 724 men, was escorted to a secure area to the south by Task Group 58.2 with forty-eight enemy aircraft in pursuit. She was then towed for temporary repairs at Ulithi by the cruiser *Pittsburgh* before being sent on for reconstruction at the New York Naval Shipyard.

The U.S. forces in the Pacific were now closing in on the Japanese home islands, and on March 22 *Wisconsin* joined *New Jersey* and *Missouri* in shelling Kutaka Shima. On March 24 a bombardment group was established under Rear Adm. John F. Shafroth that had responsibility for shelling land targets throughout the Japanese archipelago. Christened Task Group 58.7 and composed of the battleships *Wisconsin*, *Missouri*, *New Jersey*, *North Carolina*, *Washington*, *South Dakota*, *Massachusetts*, and *Indiana*, it operated until April 19. No sooner had it been created than TG 58.7 was shelling Okinawa, the floatplanes of the *Iowa*-class vessels providing spotting information for the operation. Strangely enough, it was not inside the 16-inch turrets that the worst effects of the blasts were felt, but inside the 5-inch mounts, in which seven men were hospitalized for nosebleeds and chest contractions. The psychological effect, however, was another matter, and the firepower brought to bear by TG 58.7 was such that one of *Wisconsin*'s men jumped overboard. He was recovered by a destroyer and admitted to the battleship's hospital. On April 1 came the first U.S. landings on the Japanese mainland.

Wisconsin in January 1945.
USN

The end of the *Yamato*.
USN

It was during this period that U.S. surface ships came under the most intense attack from kamikazes. Although *Wisconsin* managed to escape damage, the threat terrorized the crew and had a significant effect on discipline and morale. On April 7 the destruction of the giant battleship *Yamato*, the light cruiser *Yahagi*, and four destroyers by aircraft from Task Force 58 while TG 58.7 was lying fifty miles off Okinawa removed any possibility of the *Iowa*-class ships engaging the leading units of the Japanese fleet. In fact, much the greatest threat came from the air, and on the 11th and particularly the 12th the fleet came under sustained assault, with 151 kamikaze attacks that day causing damage to the carriers *Enterprise* (CV 6), *Intrepid* (CV 11), and *Bunker Hill* (CV 17). Five days later *Wisconsin* shot down three enemy aircraft.

On April 24, *Wisconsin* rejoined Rear Adm. Arthur W. Radford's Task Group 58.4, which consisted of her three sister ships, the carriers *Yorktown* (CV 10), *Intrepid*, *Enterprise*, and *Shangri-La* (CV 38), the escort carriers *Langley* (CVL 27) and *Independence* (CVL 22), the large cruisers *Alaska* (CB 1) and *Guam* (CB 2), the light cruisers *San Diego*, *Oakland* (CL 95), and *Flint* (CL 97), and fifteen destroyers. Operation Iceberg was in full swing, with Okinawa under withering U.S. naval and air firepower every day. On the 29th TG 58.4 again found itself under kamikaze attack, and it was not until May 11 that the crews could withdraw for rest and recuperation at Ulithi, which they reached on the 14th. The Task Group resumed battle on the 24th, and four days later the Fifth Fleet came under the command of Adm. William F. Halsey and was renamed the Third Fleet. As a result, Task Force 58 became Task Force 38.

On June 2 and 3 it was Kyūshū's turn to be pounded, after which the fleet was to refuel three hundred miles south of Okinawa. Meanwhile, a typhoon was reported to be approaching and the fleet maneuvered to avoid the threat, but it was still battered by 80-knot winds. The heavy cruiser *Pittsburgh* (CA 72) lost her bow to one-hundred-foot waves, and *Baltimore* (CA 68) and *Duluth* (CL 87) also suffered considerable damage. At the height of the storm the escort carrier *Windham Bay* (CVE 92) recorded wind speeds of 127 knots and waves of seventy-five feet. The typhoon cost the lives of an officer and five men, and thirty-three aircraft were lost. Four battleships, four carriers, six cruisers, and ten destroyers had to be withdrawn for repairs while the operation was still in progress.

Nonetheless, *Wisconsin* steamed 6,014 miles between May 28 and June 10, and on the 8th one of her floatplanes managed to rescue a downed pilot from the carrier *Shangri-La* off Kyūshū. On the 13th TF 38 reached San Pedro Bay on Leyte Gulf for a spell of well-earned rest and maintenance.

Wisconsin engaging kamikaze aircraft off Okinawa.
NARA

The fleet oiler *Cahaba* (AO 82) refueling *Wisconsin* and an *Essex*-class carrier.
USN

The U.S. fleet at sea in 1945 with two units of the *Iowa* class on the left.
USN

On July 1 TG 38.4 sailed for a bombardment mission against Honshū and Hokkaidō, which was carried out between the 13th and 15th, culminating in the destruction of refineries and industrial facilities of Nihon Steel and Wanishi Ironworks facilities at Muroran. The attack was carried out by *Wisconsin*, *Iowa*, and *Missouri* together with the light cruisers *Atlanta* (CL 104) and *Dayton* (CL 105) and two destroyer divisions. Lending their support on this occasion were the battleships *North Carolina*, *Alabama*, and HMS *King George V*. Over the course of forty minutes *Wisconsin* fired 291 16-inch shells at targets at a range of up to 31,700 yards. Between the 17th and 18th the Hitachi Miro works came under U.S. firepower, although several of *Wisconsin*'s radar sets were put out of action by blast from her own guns after the fifth salvo. A total of 833 16-inch shells were fired by the battleships during these two missions.

On the 20th the fleet undertook a record replenishment of 6,369 tons of ammunition, 1,635 tons of supplies, and 63,600 tons of fuel. Four days later the Kure naval dockyard was the target of a major air attack, which claimed 258,000 tons of shipping of all types, followed by bombing raids on Kōbe and Tokyo between the 27th and the 29th. Between August 8 and 10, TG 38.4 covered the assaults on Honshū and Hokkaidō.

On the 15th the Japanese requested surrender terms as the battleships stood by two hundred miles southeast of Honshū. A hundred men from *Wisconsin* were selected to disembark and occupy the naval base at Yokosuka, which required them to be transferred at sea to *Missouri* for landing on Japanese soil on the 29th. By the time *Wisconsin* entered Tokyo Bay on September 5, her eight months of operational service had taken her 105,831 miles and involved the refueling of 250 destroyers.

The heavy cruiser *Pittsburgh* (CA 72) minus her bow, which she lost in a typhoon on June 4, 1945.
NARA

The battleship *Arkansas* alongside *Wisconsin* at Pearl Harbor in October 1945.
USN

Wisconsin sailed from Japan for Okinawa on the 22nd, where she embarked a large number of GIs for repatriation as part of Operation Magic Carpet. After provisioning and refueling at Pearl Harbor on October 4, she sailed for San Francisco to an ecstatic welcome by Governor Walter Goodland of Wisconsin on the 15th. Eleven days later the silverware of the first *Wisconsin*, hitherto carefully preserved in the state capitol, was offered to the battleship. It was not until mid-January 1946 that she traversed the Panama Canal and anchored at Hampton Roads, joining *Missouri* in the Atlantic Fleet while *Iowa* and *New Jersey* remained in the Pacific.

Between February and May *Wisconsin* carried out numerous drills and exercises in Guantánamo Bay, while the summer found her refitting at Norfolk, with six hundred of her men being disembarked. On October 15 she sailed on a flag-showing cruise off South America as the flagship of Adm. William D. Leahy, calling at the ports of Valparaíso, Callao, Balboa, and La Guaira between November 1 and 26, during which the presidents of Chile, Peru, and Venezuela were all received on board. *Wisconsin* returned to Norfolk on November 2.

After a spell at Bayonne, *Wisconsin* spent the spring of 1947 on gunnery exercises off Central America. June 3 found her at Annapolis to embark midshipmen from the United States Naval Academy and NROTC. Together with *New Jersey*, she sailed to join Task Force 81 off Cape Henry, part of a squadron consisting of the carriers *Randolph* (CV 15) and *Kearsarge* (CV 33) and the destroyers *Cone* (DD 866), *Stribbling* (DD 867), *O'Hare* (DD 889), and *Meredith* (DD 890). After an uneventful crossing of the Atlantic during which the most significant event was underway replenishment from the oiler *Chemung* (AO 30), the squadron rounded Cape Wrath, Scotland, and reached the British naval base at Rosyth, where Adm. Richard L. Conolly, commander in chief of U.S. Naval forces in the Atlantic and the Mediterranean, hoisted his flag in *New Jersey* for the planned cruise of Northern Europe. On the 29th TF 81 sailed for Norway, the battleships making for Oslo and the rest of the squadron for Gothenburg.

On July 9, TF 81 called at Portsmouth, where the ships were visited by Admiral Sir Bruce Fraser, the victor of the Battle of North Cape. Admiral Conolly hauled down his flag on the 18th, and TF 81 turned for Guantánamo Bay, which it reached on August 1. The rest of the year passed quietly as the crew carried out the difficult task of preparing *Wisconsin* for decommissioning. In June 1948 she was assigned to the Atlantic Reserve Fleet at Norfolk and on July 1 was decommissioned in a moving ceremony.

Wisconsin in San Francisco Bay on October 15, 1945.
USN

Wisconsin exercises her main battery in February 1947.
NARA

From Kansong to Philadelphia

Wisconsin would not spend long in the reserve thanks to the outbreak of hostilities in Korea in June 1950. On the 27th of that month the United Nations adopted Resolution 83 condemning North Korean aggression. Within days it was clear that U.S. troops would be committed as the backbone of the UN forces for which heavy naval artillery would provide invaluable support.

Wisconsin was the penultimate unit of the *Iowa* class to return to service, *Missouri* having remained in commission and with *New Jersey* and *Iowa* herself emerging from mothballs in November 1950 and August 1951, respectively. *Wisconsin* left the Reserve Fleet in late 1950 and recommissioned after a much-needed refit at Portsmouth on March 3, 1951. Among the many dignitaries attending this event was the governor of Wisconsin, Walter J. Kohler.

After being provisioned with food, ammunition, and fuel, *Wisconsin* sailed on the training cruise to Guantánamo Bay, followed by steam trials with a contingent of midshipmen in the North Atlantic, with stops at Edinburgh, Lisbon, Halifax, and Cuba.

Wisconsin anchored in Cawsand Bay off the English port of Plymouth.
USN

Wisconsin being refitted prior to recommissioning.
USN

The domes protecting the 40-mm mounts removed in February 1951 as part of the refitting of the ship.
USN

U.S.S. WISCONSIN (BB-64)

RECOMMISSIONING

⚓

Saturday 3 March 1951

⚓

Norfolk Naval Shipyard
Portsmouth, Virginia

Wisconsin drydocked at Norfolk.
USN

Wisconsin's recommissioning ceremony at Norfolk on March 3, 1951.
USN

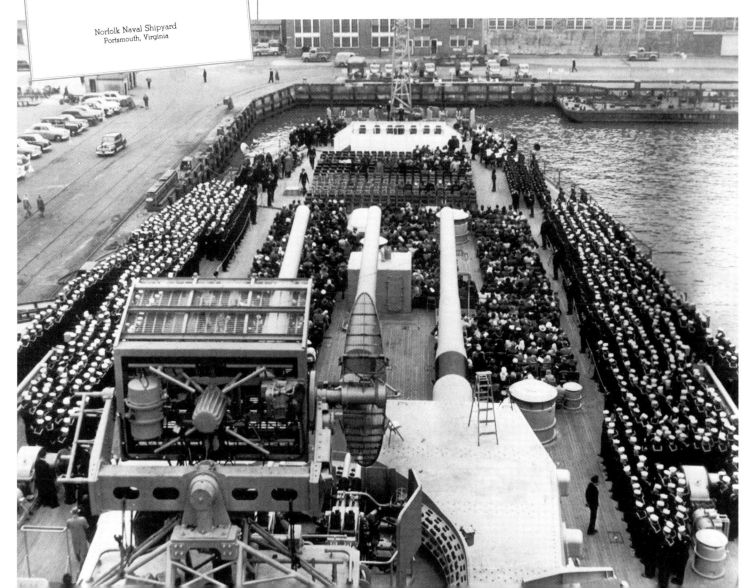

On October 25, *Wisconsin* left Norfolk for the Far East, which she reached via the Panama Canal and Pearl Harbor, arriving at Yokosuka on November 21 to relieve *New Jersey*. The commander in chief of the Seventh Fleet, Vice Adm. Harold M. Martin, immediately hoisted his flag, as did Rear Adm. Francis C. Denebrink commanding Service Force, Pacific.

On the 26th she joined Task Force 77, and her baptism of fire came December 2 in the Kansong area. Escorted by the destroyer *Wiltsie* (DD 716), her 16-inch guns provided support to the 1st Marine Division. The following day Vice Admiral Denebrink disembarked at Kangneung and on the 6th *Wisconsin* fired starshell in support of troops of the Republic of Korea (ROK). The enemy spotted, *Wisconsin* proceeded to destroy a Russian tank, two 3-inch artillery pieces, and a munitions dump. After being briefly replaced by the heavy cruiser *Saint Paul* (CA 73), *Wisconsin* resumed her bombardment operations against Kasong-Kosong and Kojo in company with the destroyer *Twining* (DD 540).

On the 12th, *Wisconsin* was visited by the commander of the 2nd Battleship Division, Rear Adm. H. Raymond Thurber, before carrying out a series of bombardment missions against fortified enemy positions over the next few days. After shelling Kojo, she turned for Sasebo, which she reached on the 16th, but she was quickly back on the gun line after rearming. On the 18th she had the unusual experience for an operational vessel of receiving on board Senator Homer Fergusson of Michigan, who departed by helicopter for the carrier *Valley Forge* (CV 45) the following day. On the 20th she participated in the bombardment of Wonsan, during which numerous suspicious vessels were sunk, and the following day her helicopter rescued a pilot from the carrier *Antietam* (CV 36). On the 28th it was the clergy's turn to visit the ship, with Cardinal Francis Spellman arriving by helicopter to celebrate mass on board. Three days later *Wisconsin* was in Yokosuka, where she remained until January 9, 1952.

In the meantime numerous prisoners of war were transferred on board for interrogation, together with wounded crewmen from the destroyers *James E. Keys* (DD 787), *Eversole* (DD 789), and *Higbee* (DD 806), who were admitted to her hospital.

On January 10 she received a visit from Korean president Syngman Rhee and his wife, but by the 14th she was providing gunfire support to the 1st Marine Division of ROK army's 1st Corps while engaged with the Chinese 45th Division. One hundred 16-inch shells had destroyed numerous mortar positions and railway lines before *Wisconsin* returned to Sasebo for rearming between January 17 and 22. On the 26th she destroyed the communications center and command post of the North Korean 15th Division while lying off Kojo. Between January 30 and February 3 *Wisconsin* lay off Wonsan to bombard a series of artillery positions at Hodo Pando, during which period the ship fired no less than 535 16-inch and 549 5-inch shells. On the 19th it was the turn of numerous mortar and fortified positions to be smashed before Vice Adm. Robert P. Briscoe relieved Admiral Martin in command of the Seventh Fleet on the 24th.

On the 25th *Wisconsin* was at Busan, where she was visited by Vice Admiral Sohn Won-yil, chief of operations of the Korean navy, U.S. ambassador to Korea John J. Muccio, and Rear Admiral Alan Scott-Montcrieff of the Royal Navy. After a brief call at Yokosuka in early March, *Wisconsin* participated in Operations Package and Derail, which consisted of interdicting enemy air traffic in the Anju area.

On March 15, while *Wisconsin* was bombarding a railway line at Songjin, a 6-inch shell fired from the Yoktaeso-Ri Peninsula destroyed one of *Wisconsin*'s starboard 40-mm mounts and penetrated the upper deck. Three men were wounded but the enemy artillery piece was silenced by salvos of 16-inch shells, after which *Wisconsin* trained her guns on a railway tunnel. A total of 152 16-inch, 449 5-inch, and 595 40-mm rounds were fired during this operation.

Wisconsin returned to Yokosuka on April 1 to find *Iowa* awaiting her. The ship then turned for Guam where she spent two days undergoing tests in floating dock *AFDB-1*. This done, she sailed for Long Beach, which she reached on the 19th after calling at Pearl Harbor. Following repairs at Norfolk she was ready to resume her role as a training vessel. On June 9, *Wisconsin* sailed from Norfolk on a midshipman cruise that took her to the Scottish port of Greenock, Brest, and Guantánamo Bay.

Wisconsin at Sasebo in January 1952.
USN

Wisconsin receiving wounded men from the cruiser *Saint Paul* and mail from the destroyer *Buck*.
USN

Wisconsin's deck penetrated by a 6-inch shell in 1952.
USN

Wisconsin entering floating dock AFDB-1 at Guam in April 1952.
USN

Between September 14 and 24 *Wisconsin* participated in Exercise Mainbrace, the first large-scale naval operation undertaken by the newly established Allied Command Atlantic (ACLANT) under the auspices of NATO. Organized on the initiative of Gen. Dwight D. Eisenhower and commanded by Adm. Lynde D. McCormick (USN), Admiral Patrick Brind (Royal Navy), and Gen. Matthew Ridgway of the U.S. Army, Mainbrace included vessels from the U.S., British, French, Canadian, Danish, Belgian, Dutch, Portuguese, and Norwegian navies. The main units aside from *Wisconsin* were the battleship HMS *Vanguard*, the carriers *Wasp* (CV 18), *Midway* (CV 41), *Franklin D. Roosevelt* (CV 42), HMS *Eagle*, and HMS *Illustrious*, the light carriers *Wright* (CVL 49), HMS *Theseus*, and HMCS *Magnificent*, the escort carriers *Salerno Bay* (CVE 110) and *Mindoro* (CVE 120), together with 5 cruisers, 96 destroyers, 33 submarines, and numerous other vessels, a total of 203 vessels, with a thousand aircraft from nine different nations operating in the Central Atlantic, Norwegian Sea, and Baltic Sea against a possible Soviet attack on Europe.

Wisconsin was subsequently drydocked at Norfolk until late January 1953 before heading for the U.S. Navy's traditional training waters, Guantánamo Bay. This was followed in early May by two weeks of exercises off Rhode Island before the ship made a three-day stopover in New York. On June 3 she sailed for Annapolis to embark midshipmen on a training cruise that took them to the Central Atlantic, with stops at Rio de Janeiro, Port of Spain, and Cuba.

After a month's refitting at Norfolk, on September 9 *Wisconsin* sailed for Japan via the Panama Canal, where she relieved *New Jersey* as flagship of the Seventh Fleet on October 14. Between late October and early December she made visits to Kōbe, Sasebo, Yokosuka, Otaru, and Nagasaki; and Christmas Day found her in Hong Kong. *Wisconsin* sailed from Yokosuka on April 1, 1954, after being replaced by the heavy cruiser *Rochester* (CA 124), returning to Norfolk on May 4 after a brief stop at Long Beach.

On June 7 she was at sea with *Iowa*, *New Jersey*, and *Missouri* to carry out various exercises off Cape Virginia, the one and only occasion on which all four units of the class sailed in company. Five days later she sailed for the North Atlantic on yet another training cruise, again taking in Greenock, Brest, and Guantánamo Bay.

In January 1955 she participated in Operation Springboard, including a stop at Port-au-Prince, before embarking on another training cruise with calls at Edinburgh and Copenhagen before returning to Cuba. In September and October she was refitted at the New York Naval Shipyard with occasional sorties to the Caribbean. Capt. Frederic Keeler briefly grounded her in the East River on October 19 but the ship lifted free after an hour. Damage was slight.

The Collision with the *Eaton* and After

On January 23, 1956, *Wisconsin* participated in another Operation Springboard with stops at Tampico in Mexico, Port-au-Prince, and the Colombian port of Cartagena.

On Sunday, May 6, a series of antiaircraft gunnery exercises was organized fifty miles off Cape Virginia with a number of political figures embarked in *Wisconsin*. The main units in the squadron aside from the battleship were the carrier *Coral Sea* (CVA 43) and the heavy cruiser *Des Moines* (CA 134), with an antisubmarine escort provided by the destroyers of the 22nd Escort Division consisting of *Bache* (DDE 470), *Beale* (DDE 471), *Murray* (DDE 576), and *Eaton* (DDE 510) in which the flotilla commander, Capt. Terrell H. W. Conner, was embarked.

The exercises were complicated by the thick fog that descended over the area, a situation that posed a particular problem for the *Eaton* (Cdr. Richard Varley), whose radar had been out of order for four days. Thanks to the budgetary restrictions then in effect, few technicians were capable of carrying out repairs at sea, and the crews were inexperienced.

At 1336 the squadron was steaming at 15 knots on a course of 180 degrees. *Murray* was leading the squadron 1,200 yards ahead of *Coral Sea*, with *Eaton* 5,500 yards off the starboard bow, *Beale* at a similar distance off the port bow, and *Bache* positioned 4,600 yards astern of the formation. At 1501 a course of 0 degrees was shaped until the carrier reported a man overboard at 1510. *Bache* was in the best position to recover the unfortunate sailor, but Conner decided to intervene with his own vessel and ordered the navigating officer to go on to 27 knots. Cdr. Varley came on the bridge shortly after and asked for clarification of what was occurring, without questioning Conner's decision. In fact, no one in the *Eaton* had the authority to intervene in the rescue without the orders of a flag officer. With *Eaton* proceeding at high speed through the fog and with no operational radar, at 1514 a huge mass of steel suddenly appeared on her starboard bow. This was *Wisconsin*, making 20 knots to reach the head of the squadron only to find her bows being crossed at 90 degrees by one of her escorts. With no time to avoid the collision, *Wisconsin* rammed *Eaton* on her starboard side abreast of her no. 1 5-inch mount. At 1518 *Wisconsin* hoisted the Delta distress flag as the two ships separated, *Eaton* coming alongside the battleship thanks to the initiative of her first lieutenant, who secured bow to stem with the anchor chain.

The collision caused severe damage to both ships. The lower section of *Wisconsin*'s bow was wrecked as far back as frame 12 with three compartments flooded, but there were no casualties. On the *Eaton*, a large opening had been made in the hull and the two forward 5-inch mounts destroyed. There was damage to the bridge and the chiefs' mess, the enlisted men's quarters had been demolished, and the ship was ten feet down by the bows. The flooding of the magazines had removed any danger of an explosion, however, and casualties were limited to bruising to Seaman Wickham's face when a crate of oranges fell on him; he was transferred to *Wisconsin*'s hospital. Both vessels were capable of making port at reduced speed, but they were met by a number of tugs sent to bring them into Norfolk Naval Shipyard, *Eaton* coming in stern first. Commander Varley remained on his bridge during the long tow and declined all food or drink, aware no doubt that his naval career was in serious jeopardy. A court of inquiry was opened that resulted in Varley and others being court-martialed and found negligent, which had serious implications for their promotion prospects. Ironically, no sailor had in fact been lost overboard from the *Coral Sea* that day. The alarm had been been raised over a seaman's cap being caught by the wind and disappearing into the sea.

Wisconsin was docked at Norfolk a week later, where her damaged bow was removed and replaced with that from her unfinished sister *Kentucky*, all 120 tons of which was conveyed from Newport News on a barge. Round-the-clock work by specialized workers succeeded in restoring the vessel in sixteen days, and *Wisconsin*, newly rechristened "Wistucky," was ready for sea on June 28 in time for her summer training cruise.

Wisconsin passing under the Manhattan Bridge on October 18, 1955.
USN

Wisconsin sailing in home waters.
USN

Damage to *Wisconsin*'s bow after colliding with the destroyer *Eaton*.
USN

Another view of *Wisconsin*'s damaged bow after the collision with the *Eaton*.
USN

Wisconsin awaiting repair at Norfolk on May 6, 1956. *New Jersey* beyond.
USN

Damage to *Eaton*'s forecastle.
USN

Loaded onto a barge with the assistance of a floating crane, the bow of the unfinished *Kentucky* en route to Norfolk for installation in *Wisconsin*.
USN

On July 9, *Wisconsin* received seven hundred midshipmen for the summer training cruise, who sailed to Barcelona, Greenock, and Guantánamo Bay, which they reached on August 24th.

Wisconsin participated in fleet exercises from late October before heading for Norfolk, where she remained until January 2, 1957. On the 15th, Rear Adm. Henry C. Crommelin commanding the 2nd Battleship Division hoisted his flag in *Wisconsin* to lead the exercises in Guantánamo Bay.

On March 27, *Wisconsin* sailed for the Mediterranean, stopping at Gibraltar on April 5. Joining Task Force 60 in the Aegean Sea to take part in Operation Red Pivot, she participated in NATO exercises in the Gulf of Xeros until turning for Naples on the 18th, where her helicopter rescued the crew of a downed aircraft from the carrier *Forrestal* (CV 59). After calling at the Spanish port of Valencia on May 10, *Wisconsin* returned to the United States on the 27th.

Between June 8 and 17 an international naval review was held at Hampton Roads to celebrate the 350th anniversary of the Jamestown colony. It was one of the largest peacetime concentrations of warships in history of the United States.

WARSHIPS ATTENDING THE HAMPTON ROADS NAVAL REVIEW OF JUNE 1957

UNITED STATES
Iowa, *Wisconsin* (battleships); *Randolph*, *Leyte*, *Franklin D. Roosevelt*, *Valley Forge*, *Saratoga* (carriers); *Albany*, *Boston*, *Canberra*, *Des Moines*, *Macon*, *Northampton* (cruisers); 24 destroyers; 5 submarines; 12 transports, LSDs & LSTs; 18 minesweepers; 10 auxiliary vessels

BELGIUM
Truffaut, *Bovesse* (minesweepers)

CANADA
Ottawa, *Assiniboine* (destroyers)

COLOMBIA
Capitán Tono (frigate)

CUBA
Antonio Maceo (frigate)

DENMARK
Holger Danske (frigate)

DOMINICAN REPUBLIC
Generalísimo (destroyer)

FRANCE
De Grasse (cruiser); *Bois Belleau* (carrier); *Dupetit-Thouars*, *Chevalier Paul* (escort destroyers); *Le Lorrain*, *Le Gascon* (fast escorts)

GREAT BRITAIN
Ark Royal (carrier); *Duchess*, *Diamond* (destroyers)

ITALY
San Giorgio (destroyer)

THE NETHERLANDS
De Zeven Provinciën (cruiser); *Groningen* (destroyer)

NORWAY
Trondheim (destroyer)

PERU
Aguirre (frigate)

PORTUGAL
Diogo Cao, *Corte Real* (destroyers)

SPAIN
Juan Sebastián Elcano (barquentine)

TURKEY
Gelibolu, *Giresun* (destroyers)

URUGUAY
Uruguay (destroyer)

VENEZUELA
Zulia, *Aragua* (destroyers)

Wisconsin in the early 1950s.
USN

Wisconsin at anchor during the international naval review at Jamestown on June 12, 1957.
USN

On June 19, 1957, *Wisconsin* sailed for South America with a contingent of midshipmen, reaching Valparaíso on July 3. She then traversed the Panama Canal to participate in Operation Strikeback with NATO forces between September 3 and 12. Together with *Iowa*, the carriers *Essex* (CV 9), *Intrepid* (CV 11), *Wasp* (CV 18), *Tarawa* (CV 40), *Forrestal* (CV 59), *Saratoga* (CV 60), HMSs *Eagle*, *Bulwark*, and *Ark Royal*, eight cruisers, fifty-one destroyers, twenty-five submarines, and five auxiliary vessels, the operation simulated the defense of Western Europe against a possible Soviet attack. Covering an area from the Atlantic to the Norwegian Sea, Strikeback was the largest operational assembly of warships since the end of World War II.

Wisconsin reached New York on November 6 and made her way to Bayonne two days later to begin the mothballing process. On March 8, 1958, she was decommissioned and placed in the reserve before being towed to Philadelphia Naval Shipyard. For the first time since 1895, the U.S. Navy had no battleship in service.

Wisconsin at sea in 1957.
USN

Wisconsin sailing from Hampton Roads to participate in Operation Strikeback in September 1957.
USN

Wisconsin in New York in November 1957.
USN

Wisconsin under tow to Philadelphia Naval Shipyard in 1962.
USN

Weeds sprout in the deck planking of *Wisconsin* at Philadelphia in 1970.
USN

Iowa, *New Jersey*, and *Wisconsin* as part of the Reserve Fleet at Philadelphia in 1962.
USN

Wisconsin and *Iowa* decommissioned in Philadelphia.
USN

Reactivation

The main instigator of the reactivation of the *Iowa* class was defense consultant Charles E. Myers, whose priority was the protection of amphibious assaults. With the assistance of two former commanding officers of *New Jersey*, Capt. Edward Snyder and Capt. Robert Peniston, in 1978 Myers wrote a forty-page report highlighting the advantages of high-performance gun platforms to support ground troops. These proposals were supported by the new Chief of Naval Operations, Thomas B. Hayward, in light of the imminent commissioning of the Soviet guided-missile cruiser *Kirov*. Although President Jimmy Carter was totally opposed to the recommissioning of the *Iowa*s, the arrival of Ronald Reagan in the White House brought on a rapid change of heart. The new Secretary of the Navy, John Lehman, successfully defended this view in Congress, arguing the advantages offered by such vessels in amphibious operations. The decision was approved by Congress in mid-1981 and *New Jersey* was recommissioned into the Navy on December 8, 1982.

Four years later it was *Wisconsin*'s turn to be reactivated. On August 8, 1986, she was towed by fourteen tugs from Philadelphia to New Orleans and was docked at Avondale Shipyards, where more than a thousand workers set to work on her. The cost of this work was estimated at $385 million, the price of an *Oliver Hazard Perry* frigate. On January 2, 1987, the ship was undocked and towed to Ingalls Shipbuilding at Pascagoula, Mississippi, where she received 33,683 yards of electrical cabling, 4,500 light bulbs, 2,000 tons of steel, 11,680 yards of tubing, 5,000 valves, and 1,500 bunks and had some 40,000 gallons of paint applied.

The modernized battleship was intended to be capable of carrying out the following functions:

1. Operating offensively as part of a carrier Task Force
2. Operating with the appropriate escort, without air cover and providing her own defense
3. Supporting troop landings
4. Conducting offensive operations against surface or land targets
5. Carrying out shore bombardment
6. Controlling a given airspace
7. Refueling helicopters of all types
8. Refueling her escort.

Source: Muir, *The Iowa Class Battleships*, 122.

The battlecruiser *Pyotr Velikiy* of the *Kirov* class, whose commissioning into the Soviet navy from the 1980s provided some of the rationale for reactivating *Iowa*-class battleships.
PRADIGNAC AND LEO

Wisconsin refitting at Ingalls Shipbuilding, Pascagoula, Mississippi, in September 1987.
USN

Wisconsin refitting at Ingalls Shipbuilding, Pascagoula, Mississippi, in September 1987.
USN

Wisconsin refitting at Ingalls Shipbuilding, Pascagoula, Mississippi, in September 1987.
USN

Wisconsin during her steam trials. Note the discone/discage antenna fitted on the forecastle.
USN

Wisconsin's recommissioning service on October 22, 1988, attended by more than 12,000 guests.
USN

The recommissioning ceremony was held on October 22, 1988, in Pascagoula, Mississippi, in the presence of the governor of Wisconsin, Thomas G. Thompson, and Texas senator John Tower, who expressed their satisfaction at the recommissioning of the ship before a 12,000-strong crowd. Commanded by Capt. Jerry M. Blesch, the ship was attached to the Atlantic Fleet as part of the Surface Action Group (SAG) formerly known as the Battleship Battle Group (BBBG) and usually composed of a battleship, a *Ticonderoga*-class cruiser, a *Kidd*- or *Arleigh Burke*–class destroyer, a *Spruance*-class destroyer, three *Oliver Hazard Perry*–class frigates, and a fleet supply ship.

After carrying out training exercises in the Atlantic and then off Puerto Rico, *Wisconsin* returned to Philadelphia for final fitting out with Tomahawk and Harpoon missiles, 20-mm Phalanx guns, and the latest detection equipment. November 29 again found *Wisconsin* off Puerto Rico, where she refueled from the oiler *Neosho* (AO 143) and replenished her magazines from the ammunition ship *Nitro* (AE 23). Throughout January 1989 *Wisconsin* was in the hands of Combat Systems Ship Qualification Testing (CSSQT) personnel, and on February 2 put in at Pascagoula to participate in the Mardi Gras celebrations.

Between the 17th and the 25th she performed numerous maneuvers and exercises in the Gulf of Mexico, with gunnery exercises carried out against Vieques Island between April 8 and 14. A Marine detachment ashore testified to the remarkable accuracy of the gunfire. After a three-day call at St. Thomas, U.S. Virgin Islands, *Wisconsin* proceeded to Norfolk for final adjustments before filling her magazines at Hampton Roads. Between June 14 and August 18, she was at Philadelphia fine-tuning her equipment. Early September found her at Ingleside, Texas, where she was visited by Secretary of the Navy Henry L. Garrett, Senator Phil Gramm, and more than 18,000 visitors, followed in December by a spell at Ingalls Shipbuilding.

The period between January 19 and March 9, 1990, was spent in training in Guantánamo Bay with a two-day stopover at Montego Bay, Jamaica, in mid-February. After embarking munitions at Earle, New Jersey, between March 10 and 13, gunnery exercises were carried out in the Caribbean in late April and early May. The summer of 1990 passed quietly, with the ship spending July at Norfolk, where members of the crewmen's families were welcomed on board on the 3rd for a brief cruise off the Virginia coast. A month later *Wisconsin* was sailing for war.

Wisconsin presents an impressive sight in 1991 despite her forty-seven years.

Wisconsin presents an impressive sight in 1991 despite her forty-seven years.
J. PRADIGNAC

A 16-inch gunnery exercise in August 1988.
USN

An RQ-2B Pioneer RPV drone maneuvered into position on the forecastle.
USN

Wisconsin astern of *Missouri* in the Indian Ocean in January 1991.
USN

The Gulf War

The invasion of Kuwait by Iraq on August 2, 1990, was immediately condemned by the United Nations Security Council, and President George H. W. Bush lost no time in authorizing the launch of Operation Desert Shield, which required U.S. and coalition naval forces to deploy to the Indian Ocean. After sailing from Norfolk on August 3, *Wisconsin* was selected to relieve *Missouri* by early September, and she covered the 8,500 miles to the Persian Gulf at a speed of 25 knots, arriving on station with her battle group on August 23. She was immediately tasked with ensuring the safety of all surface ships in that part of the Indian Ocean, and she was subsequently entrusted with protecting the passage of tankers between Kuwait and the Strait of Hormuz. On September 28, Capt. David S. Bill took command while the ship was anchored at Sitrah in Bahrain. The threat of mines did not prevent *Wisconsin* from visiting Muscat (capital of the Sultanate of Oman), Dubai, and Abu Dhabi in the United Arab Emirates, and Manama, capital of Bahrain.

With the onset of Operation Desert Storm on January 17, 1991, *Wisconsin* launched twenty-four Tomahawk cruise missiles on enemy positions over a two-day period while also coordinating surface defense as part of the Northern Persian Gulf Surface Action Group. On February 6, she replaced *Missouri* on the gun line and fired eleven 16-inch shells against Iraqi artillery positions south of Kuwait City. The following morning another twenty-nine shells were unleashed against a communications center, and that evening she was tasked with disposing of a number of suspicious vessels anchored at Khawr-al-Mufattah. Fifty heavy-caliber shells resulted in the destruction of about fifteen ships and severe damage to port infrastructure, with *Wisconsin* deploying her RQ-2B Pioneer RPV to observe the effects of her gunfire. On the 8th, she destroyed several enemy bunkers and artillery positions in the Saudi town of Ras Al Khafji.

Between the 9th and the 20th she was withdrawn from the gun line to resupply with food and ammunition, during which the ship received a visit from Senator John Warner of Virginia. On the 21st she was back, aiming fifty 16-inch shells at a command and communications position before shelling ten enemy troop barracks. On the 23rd *Wisconsin* fired ninety-four 16-inch shells in support of coalition forces, her RPV once again proving its worth in adjusting her fire as well as confirming the destruction of a number of SAM missile batteries. On the 24th and 25th *Wisconsin* pounded enemy positions in northern Kuwait and the following day anchored eleven miles off Kuwait City to serve as a relay between coalition forces. On the 27th her drone spotted two enemy ships off Faylaka Island, which were promptly eliminated by coalition aviation. On the 28th she fired her last shells against enemy positions, thereby completing her mission.

Wisconsin under way in the Persian Gulf.
USN

Wisconsin launches a Tomahawk cruise missile during Operation Desert Storm.
USN

Wisconsin's after 16-inch turret in action.
USN

Wisconsin followed by an *Iwo Jima*–class amphibious assault ship.
USN

Wisconsin fires her last 16-inch shells on May 28, 1991.
USN

During her deployment to the Persian Gulf, *Wisconsin* had steamed 45,896 miles, fired 528 16-inch, 881 5-inch, and 5,200 20-mm Phalanx shells, and 24 Tomahawk cruise missiles, and she had seen 661 helicopter landings. Her drone had spent 348 hours aloft and she had destroyed 6 mines.

On April 27, Capt. Coenraad Schroeff became *Wisconsin*'s last commander, responsible for bringing his ship and her crew safely home and preparing her for final decommissioning. Indeed, two days after he assumed command, the Chief of Naval Operations and the secretary of defense ordered the vessel to be removed from active service by the end of fiscal year 1991. As with her sisters, the collapse of the Soviet bloc and budgetary constraints left *Wisconsin* with no chance of being retained in the fleet.

On May 28, *Wisconsin* sailed from Norfolk with 186 veterans on board for a nostalgic journey to Earle, New Jersey, where she spent a week before moving on to New York to participate in the victory celebrations held in that city on June 6 in the presence of Wisconsin governor Tommy Thompson. Over the course of four days the battleship received 25,000 visitors before returning to Norfolk. On the 14th the ship played host to 1,550 relatives of the crew for a brief farewell cruise. This was *Wisconsin*'s last outing, and from the 17th her crew began preparing her for final decommissioning.

Wisconsin off New York.
USN

Wisconsin anchored off New York as part of the victory celebrations following Operation Desert Storm.
USN

The National Maritime Center of Norfolk

Wisconsin was decommissioned on September 30, 1991, and stricken from the Navy List on January 12, 1995. Nonetheless, on October 15, 1996, she was towed to Norfolk Naval Shipyard and resumed her place on the Navy List on February 12, 1998, before being shifted to Portsmouth on the opposite bank of the Elizabeth River on December 7, 2000. Although tied up at a pier adjacent to the Nauticus National Maritime Center in anticipation of being converted to a museum, under the National Defense Authorization Act of 1996, she remained the property of the U.S. Navy and was regarded as still being in reserve, with Congress imposing a number of measures to preserve her in good condition. Accordingly, the U.S. Navy paid some $2.8 million to the city of Norfolk between 2000 and 2009 to preserve the ship. On April 16, 2001, certain areas of *Wisconsin* were opened to the public as part of the Nauticus National Maritime Center. On December 14, 2009, the U.S. Navy officially transferred *Wisconsin* to the city of Norfolk, simultaneously declaring her ineligible for recall to active service. The handover ceremony took place on April 16, 2010, and on March 28, 2012, *Wisconsin* was admitted to the National Register of Historic Places.

> National Maritime Center
> 1 Waterside Dr.
> Norfolk, VA 23510
> www.nauticus.org

USS *WISCONSIN* (BB 64)

Commissioned	April 16, 1944
Decommissioned	July 1, 1948
Commissioned	March 3, 1951
Decommissioned	March 8, 1958
Commissioned	October 22, 1988
Decommissioned	September 30, 1991

Iowa and *Wisconsin* in mothballs at Philadelphia in September 1993.
USN

Wisconsin as she is today, berthed at the Norfolk Nauticus Naval Heritage Museum.
USN

Sources

National Archives and Records Administration (NARA), College Park, Md.

Naval Historical Center (NHC), Washington, D.C.

Pacific Battleship Center, San Pedro, Port of Los Angeles, Calif.

In collab., *USS* Iowa *(BB 61)*. Paducah, Ky.: Turner Publishing Company, 1996.

———. *USS* Missouri *(BB 63)*. Paducah, Ky.: Turner Publishing Company, 1998.

———. *USS* New Jersey *(BB 62)*. Paducah, Ky.: Turner Publishing Company, 1996.

———. *USS* Wisconsin *(BB 64)*. Paducah, Ky.: Turner Publishing Company, 1996.

Butler, John A. *Strike Able-Peter: The Stranding and Salvage of the USS* Missouri. Annapolis, Md.: Naval Institute Press, 1995.

Comegno, Carol. *The Battleship USS* New Jersey: *From Birth to Berth.* Battle Ground, Wash.: Pediment Publishing, 2001.

Dorr, Robert F., and Neil Leifer. *USS* New Jersey: *The Navy's Big Guns. From Mothballs to Vietnam.* Osceola, Wis.: Motorbooks International, 1988.

Friedman, Norman. *U.S. Battleships: An Illustrated Design History.* Annapolis, Md.: Naval Institute Press, 1985.

Garzke Jr., William H., and Robert O. Dulin Jr. *Battleships: United States Battleships, 1935–1992*, 2nd ed. Annapolis, Md.: Naval Institute Press, 1995.

Jordan, John. *Warships after Washington: The Development of the Five Major Fleets, 1922–1930.* Barnsley, U.K.: Seaforth Publishing, 2011.

Muir, Malcolm. *The* Iowa *Class Battleships:* Iowa, New Jersey, Missouri *and* Wisconsin. New York: Dorset Press, 1987.

Newell, Gordon, and Allan E. Smith. *The Mighty Mo. The U.S.S.* Missouri: *A Biography of the Last Battleship.* Seattle, Wash.: Superior, 1969.

Scarpaci, Wayne. *US Battleship Conversion Projects, 1942–1965: An Illustrated Technical Reference.* CreateSpace Independent Publishing, 2013.

Stillwell, Paul. *Battleship* Missouri: *An Illustrated History.* Annapolis, Md.: Naval Institute Press, 1996.

———. *Battleship* New Jersey: *An Illustrated History.* Annapolis, Md.: Naval Institute Press, 1986.

Sumrall, Robert F. Iowa *Class Battleships: Their Design, Weapons and Equipment.* Annapolis, Md.: Naval Institute Press, 1988.

Taylor, R. L. "The Commodore's Fateful Command," *Naval History* 23, no. 1 (February 2009): 56–59; available at: http://www.usswisconsin.org/wp/wp-content/uploads/2015/10/Taylor-JF-09.pdf. Accessed May 2018.

Thompson II, Charles C. *A Glimpse of Hell: The Explosion on the USS* Iowa *and Its Cover-Up.* New York: Norton, 1999.

Walkowiak, Thomas F. *Floating Drydock's USS* Missouri *BB63 2 September 1945 Plan Book.* Treasure Island, Fla.: The Floating Drydock, 1994.

Winklareth, Robert J. *Naval Shipbuilders of the World: From the Age of Sail to the Present Day.* Annapolis, Md.: Naval Institute Press, 2000.

Yarsinske, Amy Waters. *Forward for Freedom: The Story of Battleship* Wisconsin *(BB-64).* Virginia Beach, Va.: Donning Company Publishers, 2002.

Zeitlin, Richard H. *The USS* Wisconsin: *A History of Two Battleships.* Madison, Wis.: Wisconsin Historical Society Press, 2012.

Acknowledgments

I would like to express my gratitude to the following individuals for their help in the preparation of this volume: Rainer Schittenhelm, Mike Getscher, James Pobog, the late Jason W. Hall, Jack Willard, and the staff and veterans of the USS *Iowa* and USS *New Jersey*; Jacques Pradignac for making his photograph collection available to me; Josette and Patrick Caresse and Robert Dumas for their invaluable assistance in the the writing of this book. A particular word of thanks to David Way, curator of the USS *Iowa*, who has generously given of his time and knowledge at every stage in the preparation of this project and has honored me and my work by writing the foreword to it. To them all my sincere thanks.

This book is dedicated to my dear friends Bruce Taylor and Robert Dumas, without whom the *Iowa* project would have never existed.

Index

NOTE: Italicized *page references* indicate either photographs or tables.

03 Level: bridge access from, *113*; Flag Bridge, armor for, 98, *229*; Signal Shelter, viewing slit in armor protecting, *109*
04 Level Bridge, armor for, 98, *229*
05 Level, armor for, *229*

A6ME Zero. *See* Japanese aircraft
AAC Common shells, 64
Abhau, William, 329
accommodation spaces: bathtub for FDR, *165*, 300; the brig, *171*; captain's in-port cabin, *163*; enlisted men's canteen, *160–61*; enlisted men's quarters, *167*, *374*, *418*; *Iowa*, 300; *Iowa*-class, 163; officer's stateroom and bathroom, *165*; officer's wardroom, *162*, *164*; petty officers' berthing compartment, *166*; sailor's locker and berths, *168*; in *Wisconsin*'s administrative spaces, *169*
Adams, Tung Thanh, 291
advanced gun satellite launch (AGSL) system, 201
aerials: on the after stack, *121*; Mk-12, *125*; Mk-25, *126*; Satcom, 115, *120*
after main radar position, 98
Aichi B7A Grace dive bomber, 403
Ailes, John W., 192
Air Force, U.S., 201
air-conditioned living spaces, 163
aircraft carriers: battleship speeds compared with speeds of, 13; built at New York Naval Shipyard, 15; Washington Naval Conference agreement on specifications for, 4. *See also specific carriers*
air-search radar units, 114
Akagi (Japan), 16
Akagi-class battlecruisers (Japan), 3
Alabama: at Battleship Park, Mobile, Alabama, *17*; Pacific War, *245*, 249, 321, 403, 466, 473; questioning hole in *Iowa*'s deck, 244; as *South Dakota*–class battleship, 13
Alaska, 253
Almond, Edward M., 428
altitude reference assembly (ARA), for Harpoon missiles, 115
Amagi (Japan), 470
America (carrier), 275, 352
Ammonoosuc, 202

anchor chains, 149, *150*
anchors, 149, *149–50*
AN/SRA-57 and AN/SRA-58 aerials, 115
antiaircraft armaments: characteristics, *71*; exercise during Pacific War, *395*; 40-mm Bofors Mk-2 56-caliber guns, 68, *68–70*, 73, 114, 123; *Iowa*'s, *231*; *Missouri*'s, *437*; *New Jersey*'s, 73; 20-mm 70-caliber Mk-4 guns, 71, *71–72*, 73
AN/WCS-3 transmitters/receivers, 115
AP Mk-8 (armor-piercing) shells, 46
AP tracer shells, 71
APDS Mk-149 shells, 74
Arizona, 8, 15
Arkansas, 473
armaments: antiaircraft, 68, *68–70*, 71, *71–73*; Harpoon system, 78, *78–79*; inflatable targets for gunnery exercises (1980s), *62*; main 16-inch, *46–47*, *48*, *58–61*; *Missouri*'s, *46–47*, *80–81*; original and modifications to, 46, *46*; Phalanx CIWS, 74, *74–75*; secondary 5-inch, 64, *64–67*; self-defense, 79, *79*; Tomahawk system, 76, *76–77*; Washington Naval Conference agreement on maximum caliber for, 4. *See also* antiaircraft armaments; 5-inch guns; 40-mm Bofors Mk-2 56-caliber guns; powder; shells; 16-inch guns; turrets
armor, conning tower, 98, *99*, *229*
armored box launchers (ABLs), for Tomahawk missiles, 76, *76*, 373, *437*, *442*
Arnold, Henry H. "Hap," 234
Ascendant, Operation, 430
Ashcroft, John, 437
Atkeson, John, 194
Atlantic Ocean: *Iowa* during World War II in, *231*, 233, *234–36*, *234–35*; *Iowa* with NATO in, 261, 264, 284; *Iowa*-class battleships off Virginia Capes, 263, *346–47*, 347, 434, 480; *Missouri* exercises in, 415, 431; *New Jersey* and *Wisconsin* exercises in, 348; *New Jersey* during World War II in, 316; *New Jersey* in, 336; *New Jersey* steam trials near Venezuela, 312; *William D. Porter* during World War II in, 235, *236*; *Wisconsin* joins *Missouri* in, 474; *Wisconsin* training cruise in, 475, 498; *Wisconsin* with NATO in, 480, 488
Atlas D CGM-16 rockets, 201

Austria, German annexation of, 13
aviation technicians, *143*. *See also* embarked aviation

Babcock & Wilcox Type-M boilers, 82, *82*
Backherms, Robert Wallace, 291
Badger, Oscar C., 253, 327, 338
bakery, *158*
Baltimore, 245, 394, 399, 466, 471
Baltops 85, Operation, 63, 275
band, ship's, *173*
barbershops, 163, *419*
Bard, Ralph A., 457
basketball, shipboard, *175*
battery plotting rooms: rangefinder data sent to, 123, *130*. *See also* plotting rooms
Battle, Dwayne Collier, 291
battle efficiency award markings, *199*
battle honors, 198, *198–99*
battlecruisers: Japanese, 3, *4*, *5*; Washington Naval Conference agreement on specifications for, *4*
Battleship (movie), 439
Battleship *Arizona* Memorial, 454
Battleship Design Advisory Board, 14
battleship designs: naval architects working on, *12*; Washington Naval Conference agreement and, 3, 11–14, *14*
Battleship *Missouri* Memorial, 455
Battleship *New Jersey* Museum, 355, 371, 373, *374–77*
battleships: modern naval warfare and, 266–67, 271, 275, 283–84; Washington Naval Conference agreement on specifications for, 4
Bayonne Naval Shipyard, 228
Beary, Donald B., 312
Beaufort, 437
Becton, Julian, *192*
Bell, 238, 245
Bell H-13 helicopter, 133
Bell UH-1 Iroquois Huey, 133
Bennington, 15, 328
Bethlehem Steel, 23
BGM-109 Tomahawk cruise missiles, 46, 76
Bill, David S., *195*
Biloxi, 245, 335, 466

Black paint, 184
Blakey, Walter Scot, *291*
Blamey, Thomas, 406
Blesch, Jerry M., *195*, 498
Block Island, 236
boat boom, after port, *155*
boat boom, after port, on *Iowa*, 154
boat cradle, port, on *Iowa*, 154
Boeing Integrated Defense Systems, 78
Boeing Vertol CH-46 Sea Knight, 133, *141*
boilers: Babcock & Wilcox Type-M, 82, *82*; feed water analysis, *92*; fireman operating, *86*
Bold Mariner, Operation, 353
bolsters or bolster plates, 149
Bopp, Peter Edward, *291*
Boston, 245, 394, *407*, 468, *486*
Boston Metals Company, 209
Bradford, 238, 317
Bradley, Omar N., 341
Bradshaw, Ramon Jarel, *291*
Bremen, 75
bridge: access from 03 Level to, *113*; interior view, *106–7*; *Iowa*'s, 98, *99–104*, 251; shaft revolution indicator on, *105*; shooting the sun on, *109*; stages of, 98
brig, *171*
Brind, Patrick, 480
Briscoe, Robert P., 478
Brodie, Robert, *194*
Brooklyn, 236
Brooks, Charles, *194*
Brown, 238, 245
Brown, William, *194*, 420, 422, 426
Bruton, Henry C., *195*
Bryant, Carleton F., 231
Bryson, William, *192*
Buch, Phillip Edward, *291*
Bunker Hill, 238, 240, *245*, 324, 328
buoyant life nets, on *Iowa*, 153
Bureau of Construction and Repair, U.S. Navy: cruiser killer concept of, 13–14; *Iowa*-class designs and, 14
Bureau of Ships (BuShips), 200
Burns, 238, *245*, 317
Burrowes, Thomas, *195*
Bush, George H. W., 267, *271*, *291*, 446, 452
Bush, George W. and Laura, 454

Cabot, *245*, 249, 251, 324, 466
calisthenics, *174*
Callaghan, William, *194*, 382
Camden, 209
camouflage, 184
Canberra, 245, 251, *486*
canteens, enlisted men's, *160–61*
capstans, *151*
Carey, Wes, 452
Caribbean: BLASTEX 1-87, 284; exercises, 261, 336, 347, 481, 499; *Missouri*'s first cruise in, 426; *New Jersey*'s return to, 341; U.S. presence in, 355; *Wisconsin*'s trials in, 466
Carley floats (rafts), on *Iowa*, 153
Carney, James, *194*
Carney, Robert B., 406
Casey, Eric Ellis, *291*
Casey, Sillas, III, 205
catapults: helicopter landing pad replacing, *137*; OS2U Kingfisher floatplane on, *395*; rocket-assisted, for remotely piloted vehicles, *133*, *133*; Type-P Mk-6, 133, *133*

Catchpole, Operation, 238
Chantry, Allan J., 13
charge: for 5-inch guns, 64; for 16-inch guns, 47
Charleston Naval Ordnance Plant, 46
Charrette, 238, *245*, 317
chart house, 98, *110*
Cher, music video aboard *Missouri* by, 439
Chernesky, John, *194*
Chicago, 13, *407*
Chitose (Japan), 245
Chiyoda (Japan), 245
Chromite, Operation, 428
CHT (collection, holding, and transfer) sewage system, 163
Cimarron, 247
CIWS (close-in weapon system): Phalanx Mk-15 Block 0, *46*, 74, *74–75*; radars for, 114
Clark, Joel B., 382
Clark, Joseph J., 320, 342
Clark, Mark, 261, 434
Claxton, 251
Cleveland-class light cruisers, 200
Coast Battleship No. 4, 202
cobblers, 163
Cogswell, 234
cold stores, 158, *158*
Colombia's Thousand Days' War, 205
Colwell, John B., *194*
command and landing support ships, 200
command position, 83
command transmitters, 98
commanding officers: *Iowa*, *192–93*; *Missouri*, *194*; *New Jersey*, *194*; *Wisconsin*, *195*
communications equipment, 115, *119*, *120*, 130
compass binnacle, *110*
Connecticut, 15
Conner, 238, *245*
Conner, Terrell H. W., 481
Conolly, Richard L., 336–37, 474
control tower (conning tower): 03 Level Signal Shelter viewing slit, *109*; bridge interior view, *106–7*; chart room, *110*; compass binnacle, *110*; flag deck, *112*, *113*; inside, during Pacific campaign, *108*; *Iowa* class, 98, *99–104*; meteorologists on fantail (1985), *111*; rear passage along, *105*; signal locker, *112*
Cooper, Joshua, *192*, 261
Cooper, Philip H., 205
Cosgrave, Lawrence Moore, 406
Cowell, 238, *245*
Cowpens, 238, 240, *245*, *407*, 466
Cramer, John Peter, *291*
cranes, for embarked aircraft, 133, *133*, *134*, *137*
crew entertainment, *173*, *175*, *176*, *177*
Crommelin, Henry C., 486
cruisers, Washington Naval Conference agreement on, 4
Cunningham, Andrew, 236
Curtiss SC-1 Seahawk reconnaissance floatplanes, 133, *136*

Dalton, John H., 454
Damage Control Central (DCC), 83
Davison, Ralph E., 394, 395, 468
Dawson, Lawton, 235
dazzle design, 184, *185*, *189*, *190*
DC/PS (digital computer power supply), for Harpoon missiles, 115
Deck Blue, 20-B paint, 184
deck winches, 153, 154

Decoy, Operation, 261
Delaware-class battleships, 3, 6
Denebrink, Francis C., 478
Denfeld, Louis E., 420
Dennison, Robert, *194*
dental services, 180, *181*
Derevyanko, Kuzma Nikolaevich, 406, *409*
desalinating plants, 82
Desecrate I, Operation, 244, *245*, *246*, 319–20
Desert Shield/Desert Storm operations, 446, *446*, *450*, 505, *506*
design power, 82
Devaul, Milton Francis, Jr., *291*
Dewey, Thomas E., 411, 412
diesel generators, 82, *82*
Dobson, Bennett M., *192*
Draemel, Milo F., 457
Dreadnought, 3
drinking water, 158
Driscoll, Alfred, *338*
drones: Iraqi soldiers' surrender to, 446; *Missouri* refitting and, 437; recovery of, *145*; tactical, 133; *Wisconsin*'s, Gulf War and, 505, 511. *See also* RQ-2B Pioneer remotely piloted vehicle
DSMAC (Digital Scene Matching Area Correlator), for Tomahawk missiles, 114
Duke, Irving T., *194*, 426
Duluth, 399, 471
Dunning, Allan, 304

Earnest Will, Operation, 284, 439
Eaton, 209, 481, *485*
Edison, Carolyn, 304
Edsall, Warner, *194*, 434, *434*
Eisenhower, Dwight D., 480
electric motors, 82
electrical cabling, 83
embarked aviation: aviation technicians, *143*; Curtiss SC-1 Seahawk, *136*; helicopter landing pad, *137*; Kaman SH-2 Seasprite on, *139*; MK-6 catapults and handling crane, *133*; preflight briefing, *140*; RQ-2B Pioneer, *133*, *144–45*; Sea Knight on *Iowa*, *141–42*; Sikorsky HO3S-1 Horse, *138*; Sikorsky SH-60B Seahawk, *142*; types and characteristics of, 133; Vought OS2U Kingfisher, *134–35*
emergency life rafts, inflatable, on *Iowa*, 154
engine plant: access to spaces for, *83*; components, *82*; evaporating equipment, *83*; fireman operating a boiler, *83*; interior view, *84–85*; on *Missouri* and *New Jersey*, 87; technical documentation, *92*; weight of, 83
engines: capstan, *152*; General Electric electro-hydraulic, 149
enlisted men: canteen, *160–61*; design complement for, 163; galley, *158*; quarters, *167*, *374*, *418*
Enterprise: battleship speeds compared with speed of, 13; fleet exercises (1989) and, *370*; kamikaze attacks, 399, 471; Navy Day (1945), *413*; Operation Desecrate I and, *245*; Pacific War, 394, 395
Entwistle, Frederick, *192*
Ertegün, Münir, 412
Escalante, 234
Essex: kamikaze attacks, 399; NATO and, 488; Pacific War, 240, 249, 328, 403, 466
Essex-class carriers, 15, 16, *329*, *472*
evaporating equipment, section of, *86*
Everhart, Leslie Allen, Jr., *291*

fantail: calisthenics on, *174*; drones on, *144*, *145*, 437; floatplanes, then helicopters on, 133, *137*, *138*, *141*, *314*, *344*, 349; meteorologists on, *111*; scrubbing, on *Iowa*, *178*; snowman on, *179*; whaleboats on, 154
Fazio, Tony, 235
Feinstein, Dianne, *194*, 437
Fergusson, Homer, 478
fire control: 05 Level Station, *229*; for 5-inch battery, 123; *Iowa* vs. *Yamato*, 249; position, 98, *123*; radars, 114
fire-control rooms (firerooms), 83, 87, 123, *248*
Fisk, Gary John, *291*
5-inch guns: characteristics and protection of, 64, *64*; on *Iowa*, *65*, *67*; maintenance on, *65*; Mk-37 rangefinders for, 123; mount interior for, and crewmembers, *66*; mounts on *Iowa* for, *231*; opening fire with, *67*; radars for, 114; shell casings from, *66*
flag deck, *112*, *113*
Flaherty, Michael, *195*
Flatley, *197*
fleet replenishment ships, 200, 201
Flintlock, Operation, 238
floating dock *ABSD-2*, 251, *251*, *252*
floatplanes: as embarked aircraft, 133. *See also* catapults; Curtiss SC-1 Seahawk reconnaissance floatplanes; Vought OS2U Kingfisher reconnaissance and antisubmarine floatplanes
Florida, 6
Florida-class battleships, 3
Fogarty, William, *194*
Foley, Robert J., *195*
Foley, Tyrone Dwayne, *291*
food provisions, 158, *158–62*, *416*, *417*
Forager, Operation, 245, 323
Fore River Shipyard, Quincy, Massachusetts, 204
forecastle: equipment on, *151*; *Iowa's*, *295*; mooring lines on, *148*; scrubbing, *239*; shells on, *53*; teak after seventy-two years on *New Jersey*, *24*
Forest, Francis, 304
Forrestal, James, 382
40-mm Bofors Mk-2 56-caliber guns, 68, *68–70*; radars for, 114; rangefinders for, 123
France, Washington Naval Conference agreement and, 3–4, *4*
Franklin D. Roosevelt, 480, 486
Franks, 328, *329*
Fraser, Bruce, 337, 406, 474
fuel consumption, 82
fuel hoses, gantry on starboard side for, 83, *91*
Fusō-class battlecruisers (Japan), 3

galleys: enlisted men's, *158*; officers' wardroom, *162*
Garrett, Henry L., 499
Gedeon, Robert James, III, *291*
"geedunk," 169
Gendron, Brian Wayne, *291*
General Board of the United States Navy, 11, 14
General Electric electro-hydraulic engines, 149
General Electric reduction gear, 82, *82*
Geneva Naval Conference (1927), 11
Germany, Austria annexed by, 13
Ghilarducci, Randy, 353
Gibbs & Cox, *Iowa*-class designs by, 14
Glenn, Walter, *194*
Gneckow, Gerald, *192*, 267
Goins, John Leonard, *291*
Goldsborough, Patrick, *195*

Goodland, Mrs. Walter S., 457, *461*
Goodland, Walter, 474
Gramm, Phil, 499
Gratitude, Operation, 468
Great Britain, Washington Naval Conference agreement and, 3–4, *4*
Great White Fleet, 204, 205, 439
Green, Clark L., *195*
ground tackle: anchors, *149*; types and characteristics, 149
Guam, 253
guided-missile battleship (BBG), 200, 209
guided-missile monitor battleship (BB MG), 200
Gulf of Mexico, 284, 499
gyrocompass repeater, 98
gyroscope, Mk-41, in plotting room, 130

Hailstone, Operation, 240, 317
Hall, 236
Halligan, 236
Halsey, William F.: Battle of Leyte Gulf and, 324; crew morale on *New Jersey* and, 327, *327*; on *Iowa* refitting in Admiralty Islands, 251; Japanese surrender and, 335, 406; joins Task Group 38.2, 247; Kingfisher ditching off *New Jersey* and, 323; Kurita pursuit with *New Jersey* and, 249; *Missouri* as Fifth Fleet flagship and, 399; *Missouri* launching and, 382; *New Jersey* recommissioning and, *338*; previous *Missouri* flareback incident and, 205; shifts flag to *South Dakota*, 412; Task Force 38 and, 471; *Wisconsin* assigned to, 466
"Halsey's Typhoon," 251, 327
Hampton Roads Naval Review (1957), 263, 486, *486*, *487*
Hancock, 251, 324, 466
Hanson, David L., *291*
Hanson, Edward W., 468
Hanyecz, Ernest Edward, *291*
Harpoon missiles (system): guidance system, 115; *Iowa* fires, *286*; *Iowa* modernization and, 267; *Iowa's* launchers for, 77, *79*; *New Jersey* refitting (1981), 355; schematic drawing and examples, *78*; specifications, *78*; *Wisconsin* refitting (1988), 498
Hart, Thomas C., 13
Hartwig, Clayton Michael, *291*
hawse pipes, 149
hawsehole, painting the, *149*
hawsers, 149; splicing, *179*
Hayward, Thomas B., 492
Haze Gray, 5-H paint, 184
HC Mk-13 (high-capacity) shells, 46
HE shells, 71
HE/HC shells, 64
Helfrich, Conrad Emil Lambert, 406
helicopters: as embarked aircraft, 133, *138–39*, *141–43*; fantail landing pad, *137*, *344–45*, 349; refitting and conversion projects and, 200
Helton, Michael William, *291*
Higgins, John M., *195*
high-altitude search radar, 114
Hill, Tom, *194*
Hillenkoetter, Roscoe, *194*
Hiroshima, 254, 403
hoisting winch, 154
Holden, Carl, *194*, 304, 312
Holloway, James L., *192*, 431
Holt, Scott Alan, *291*
Hope, Bob, 353, 361, 430
Hopkins, Harry, 234

Hornet, 245, 328
hospital, on-board, 180
Housatonic, 234
Hovgaard, William, 14
Hsu Yung-Ch'ang, 406
Huffman, Leon, *194*
Hull, 251, 327
Hunters Point Shipyard, 251, *252*, *259*
hurricanes, 428
Hustvedt, Olaf M., 238, 243, 247, 316
Hyuga (Japan), 245, 470

Iceberg, Operation, 328, 395, 471
Idaho, 15, *407*
Illinois, 15, 209, 213
Illinois-class battleships, 205
Independence-class escort carriers, 200, 324, 466
Indiana: *Iowa* and, 237–38; Pacific War, 245, 249, 321, 323, 328, 398, 470; as *South Dakota*–class battleship, 13, *13*
Indianapolis, 245, 317, 328
Ingalls Shipbuilding, Pascagoula, Mississippi, 267, *268*, 492, *494–96*, 499
Ingersoll, Royal E., 312
Intrepid: kamikaze attacks, 251, *325*, 471; NATO and, 488; Pacific War, 240, 324, 328, 395, 399
Iowa: administrative offices, *172*; after superstructure (1944), *48–49*; antiaircraft guns, 68; approval of, 15; armaments and armament modifications, 46–47, *46*; battle honors, *198*; boats aboard, 153–54; bow view, *30*; "Broadway" on, *24*; commanding officers, *192–93*; commissioning, 228; construction, *220–24*, 227, 228, *229*, *230–31*; damaged by submerged rock or wreck, 231; decommissioning, 254, 264, 265, 491, 512; design complement, *163*; dimensions, displacement, and protection, 19; discone/discage antenna, *121*; in dry dock at Norfolk, *276–77*; in dry dock for repairs, 251, *251–52*; enlisted men's accommodations, 158, 161, 167; explosion in turret 2 (1989), 288–89, *289–93*, 294; fireroom, *248*; firing main battery, *63*; as first in its class, 16–17; 5-inch Mk-28 battery, *65*; 40-mm Bofors mounts in, *69*; forward superstructure of, *99*; Gulf of Mexico and Middle East operations, 284; gunnery exercises, *196*, 271, *271–72*; at Hampton Roads Naval Review (1957), *486*; Harpoon system in, *78*; Japanese surrender and, 254; Kaman H-2 Seasprite on, *139*; Korean War, 258, *258–62*, 261, *433*; launching, *225–26*; main rangefinder atop foremast, *123*; maneuverability, 82; Marshall Islands campaign, 243, *243*; metacentric height, 19; Mk-25 aerial, *126*; Mk-37 rangefinder, *126*, *127*; modernization of, 266–67, *266–70*; as museum, 210, 300, *300–303*; NATO and, 261, 263–64, *275*, *279–81*, 284; in 1950s, *128–29*; in 1944, 18–19, *28–29*, *50–51*; in 1946 and 1947, *255–57*; opening fire with secondary battery, *67*; operational photos, *296–99*; Pacific War, 237–39, 241–50, 254; paint and camouflage, 185, *188–89*; Phalanx mount on, *74*, *75*; plans for, *27*; previous ships named, 202, *203*; radars, 115, *117*; radio call signals, 115; rear 16-inch turret (1943), *56–57*; refitting of, 251, *251–52*; refueling, *253*, *278*, *287*; religious service in, 180, *253*; remotely piloted vehicle on, 133; rolling in high seas, 83; Roosevelt to Tehran Conference on, *233*, 234–36, *237*; RQ-2B Pioneer drone

on, *144*; scrubbing fantail on, *178*; Sea Knight on, *141–42*; service career, 219–303; shells, handling on board, 54; shells awaiting stowage, *282*; ship's store, *173*; 16-inch guns, *247*; Sky 1, 2, 3, and 4, 123, *127*; spare parts stores, *172*; speed of, 82; SRBOC launchers on, 75; stereoscopic Mk-48 GFCS Mods 1–5 in, 123; stern view, *31*; Task Force 38 and, 253; Task Force 58 and, 244–45, *245–46*; Task Group 38.2 and, *247*; Task Group 58.3 and, 238; Task Group 58.4 and, 247; Task Unit 28.1.1 and, *196*; Truk campaign, 240, *241*, 242; 20-mm guns, *72*, *242*; Woodward laying keel of, *20*; *Yamato* compared with, 249
Iowa-class battleships: armor belts, 23, *23*; armor deck hatch, *25*; body plan, *20*; builders of, 15–16; decks, 23; designs for, 14, 46; genesis of, 3–11; never completed, 206–8, *209*; plans for, *27*, *32–33*, *40–45*; preservation and legal status, 210–16; profile views, *38–39*; refitting and conversion projects, 200–201; sailing in company, 261, 263, *264*; technical characteristics, 18–45; off Virginia Capes (1954), *263*, *346–47*, 347, 434, 480; Washington Naval Conference agreement and, 11–15; as wet ships forward, 83. *See also* shipboard equipment
Ise (Japan), 245, 470
Ise-class battlecruisers (Japan), 3
Isitt, Leonard M., 406
Isuzu (Japan), 245
Italia (ex-*Littorio*), 236
Italy, Washington Naval Conference agreement and, 3–4, *4*
Itō Seiichi, 328
Izard, 238, *245*, 317

Jamboree, Operation, 328, 468
James, R. K., 251
Japan: bombardment against, 253–54, 328, 398, 403, 468, 470, 473; China invaded by, 13; surrender, 254, 406, *408–11*, 415, 473; Washington Naval Conference agreement and, 3–4, *4*
Japanese aircraft: Pacific War, 323, 324, 328, 394, *395*, *397–98*, 399. *See also* kamikaze attacks
Jennings, William W., *192*
Johnson, Louis A., 420, 426
Johnson, Reginald L., Jr., *291*
Jones, Brian Robert, *291*
Jones, Nathaniel Clifford, Jr., *291*
Joy, C. Turner, 340
Jupiter SM-78S weapon, 200
Justice, Michael Shannon, *291*

Kaiss, Albert Lee, *194*, 437, 446, 453–54
Kaiyo (Japan), 470
Kaman SH-2 Seasprite, 133, *139*
kamikaze attacks, 249, 251, 325–26, 398, *398*, 468, 470, 471, *471*
Katori (Japan), 240, *241*, 317
Katz, Douglas, *194*
Kawasaki Ki-61 Hien Tony. *See* Japanese aircraft
Keeler, Frederic, *195*, 481
Keith, Robert, *194*
Kennedy, John F., 349
Kentucky: approval and fate of, 15, 209; bow of, for *Wisconsin*, 481, *485*; construction, 206–8; conversion plans for, 200; minus her bow, *211*; towed for demolition, *212*
Killen, 251
Kimble, Edward J., *291*

King, Ernest, 234
King II, Operation, 247, 324
Kirov (USSR), 266
Knox, Frank, 15
Kohler, Walter J., 475
Kongō-class battlecruisers (Japan), 3, *4*
Korean War: Armistice Line (1953), *380*; *Iowa*, 154, 216, 258, *258–62*, 261; *Missouri*, 428, 429–30, 430–31; *New Jersey*, 216, 339, 340–42, *340–46*; *Wisconsin*, 341, 342, 475, *475–77*, 478
Kurita Takeo, 247, 248, 249, 324, *325*

La Follette, Robert M., Jr., 457
Lackawanna, 247
Langley, 245, 249, 328, 466, 471
laundry, 163, *170–71*
Lawrence, Richard E., *291*
LCM6 landing craft, 200
League Island, 16
Leahy, William D., 234, 411, 474
Leclerc de Hautecloque, Philippe, 406
Lee, Willis A., 243, 319, 323
Lehman, John, 82, 266, 267, 283, 355, 437, 492
Leverton, Joseph, *194*
Lewis, Richard John, *291*
Lexington: battleship speeds compared with speed of, 13; converted to aircraft carrier, 4, *10*; *Iowa* training sorties with, 243; *New Jersey* floatplane collecting aviators from, 323; Pacific War, 245, 249, 319, 394, 466, 468
Leyte Gulf, Battle of, 248, 324, 466, 486
library, shipboard, *177*
Light Gray 5-L paint, 184
lighting systems, 83
living spaces, 163. *See also* accommodation spaces
London Naval Conference (1930), 11
Loud, Wayne, *192*, *193*
Louisville, 245, 466
Love III, Operation, 468
Luken Steel, 23
Luzon, Battle of, 249

MacArthur, Douglas, 244, 406, *408*, 428
machine shop, *90*
Macomb, 236
Magana, Álvaro, 360
Magic Carpet, Operation, 254, 336, 411, 474
Maikaze (Japan), 240, *240*, 317
mail, receipt of, *176*
main fire-control position, 98
main 16-inch armaments. *See* 16-inch guns
Maine, 15
Maine-class battleships, 205
Manila Bay, 249, 468. *See also* Philippines, Pacific War and
Marianas campaign, 245, 323
Mariner, Operation, 261
Marshall, George C., 234
Marshall Islands campaign, 238, *239*, 240, *241–43*, 243–45
Martin, Harold M., 340, 341, 478
Martinez, Jose Luis, Jr., *291*
Massachusetts: Pacific War, 245, 249, 321, 328, 398, 466, 470; as *South Dakota*–class battleship, 13, *13*
McCain, John S., 466
McCann, Allan R., *192*, 247
McCarthy, Joseph R., *195*
McCorkle, Francis, *194*
McCormick, Lynde D., 480

McCrea, John L.: on hole in *Iowa*'s deck, 244; as *Iowa* commanding officer, *192*, *193*; *Iowa* commissioning and, 219, 267; *Iowa* striking submerged rock or wreck and, 231; Roosevelt to Tehran Conference and, 235; uninterrupted time on bridge for, 247; Victory (dog), 234, *234*
McDonald, David L., 349
McLean, Heber H., 336
McMullen, Todd Christopher, *291*
media and audiovisual room, shipboard, *176*
medical department, 180, *180–83*
Mediterranean Sea: *Iowa* during World War II in, 236; *Iowa* in, 263, 284, 294; *Missouri* in, 412, 420; *New Jersey* in, 348, 361; *Wisconsin* in, 486
Melson, Charles, *194*
Menocal, George, *194*
Merida, 202
meteorologists, *111*
Metten, John, 14
Miami, 249, 324, 466
Miller, Todd Edward, *291*
Milligan, Richard D., *194*, 293, 361
Miner, John O., *195*
Minneapolis, 16, 238, 240, 317, 329
missile guidance unit (MGU), for Harpoon missiles, 115
Missouri: anchor, *150*; approval of, 15; armaments, *46–47*, *80–81*; battle honors, *198*; bow view, *36*; bridge, *104*, *106–7*, *109*; capstans, *151*; catapult launch of OS2U Kingfisher from, *135*; commanding officers, *194*; commissioning, 382, *390*, 455; construction, 22, *383–85*, *387–90*; crew quarters, *168*; crossing the line in, *177*; decommissioning, 434, *436*, 455; design complement, *163*; dimensions, displacement, and protection, *19*; engine room, *87*; fitting out at New York Naval Shipyard, 15; fuel consumption during Okinawa and Kyūshū operations, 82; hull of, dry dock view, *25*; infirmary, *180*; Japanese aircraft attack, *397–98*, 398; Japanese surrender and, 254, 406, *408–11*, 415; Korean War, 428, *429–34*, 430–31, 434; launching, *386*; life aboard, *416–19*; Mk-37 rangefinder and Mk-12 aerial on, *125*; Mk-38 Mods 6–7 rangefinders in, 123; movies filmed aboard, 439; as museum, 453–54, *453–56*; Pacific War, *396–412*, 398–99, 403, 406; paint and camouflage, *185*, *190–91*; plans for, *32–33*; previous ship named, 204–5, 439; propellers, *95*; radars, *117*; radio call signals, 115; Reagan and Bush administration service, 437, *437–51*, 439, 446; reduction gear for, 82; rudders, *95*; searchlight, *146*; service career, 382–456; shakedown cruise, *391–93*; stern view, *37*; stranding on Thimble Shoals, 420, *420–27*, 422, 424, 426; Task Force 58 and, 394; Task Group 38.4 and, 253; Task Group 58.4 and, 395; Task Group 58.7 and, 470; 33-foot boat on (1944), *153*; trials in Chesapeake Bay, 394; upper control tower, *100–101*
Missouri Memorial Association, 454
Mitchell, William "Billy," 204
Mitscher, Marc A., 238, 316, 321, 394, 414, 468
Mitsubishi G4M Betty torpedo bomber. *See* Japanese aircraft
Mk-2 boats, 33-foot, 153, *154*, *155*
Mk-2 officer's motor barge (OMB), 154
Mk-3 40-foot launches, 153, *156*

Mk-3 utility boats, 154
Mk-4 radar, 114, 123
Mk-6 life rafts, inflatable, 154
Mk-6 rafts, 153, *157*
Mk-8 electromechanical rangekeeper, 130, *131*
Mk-8 radar, 114, *132*
Mk-10 26-foot boats, 153
Mk-12/22 radar, 114, 123, *125*, *126*
Mk-13 FC radar, 130
Mk-13 radar, 114, *115*
Mk-14 fire control unit, *68*
Mk-19 radar, 114
Mk-23 Katie shells, 46
Mk-25 radar, 114, 123, *126*
Mk-29 radar, 114
Mk-35 radar, 114
Mk-36 SRBOC launchers, *46*
Mk-37 directors, 146
Mk-37 rangefinders, 123, *125*, *126*
Mk-38 directors, 123, *390*
Mk-38 Mods 6–7 rangefinders, 123
Mk-41 gyroscope, 130, *132*
Mk-42 rangefinders, 123
Mk-45 rangefinders, *124*
Mk-48 calculator, 130
Mk-48 GFCS Mods 1–5 rangefinders, stereoscopic, 123
Mk-49 rangefinders, 123
Mk-51 director, *132*
Mk-51 rangefinders, 123
Mk-52 coincidence rangefinders, 123, *124*
Mk-53 coincidence rangefinders, 123
Mk-56 rangefinders, 123
Mk-57 rangefinders, 123
Mk-63 rangefinders, 123
Mk-66 telescopes, 123
Monaghan, 245, 251, 327
Monitor, 15
Monterey, 238, 245, 312, 316, 466
Montgomery, Alfred E., 319
Moorse, John, *192*, 294
Moosally, Fred, *192*, 284, *288*, 291
Morris, Frank G., 420, 422, 426
Morrison, Robert Kenneth, *291*
Moses, Otis Levance, *291*
motor launches, 40-foot, 153
motor launches, 50-foot, 153
motorboats, 35-foot, 153
Ms 13—Haze Gray (paint) System, 184, *185*
Ms 21—Navy Blue (paint) System, 184, *185*
Ms 22—Graded (paint) System, 184, *185*, *188*
Ms 32-1B—Medium Pattern System, 184, *185*, *189*
Ms 32/22D—Medium Pattern System, 184, *185*, *190*
Muccio, John J., 430, 478
Murray, Stuart, *194*
Musashi (Japan), *248*, 249, 324, *324*
museums: *Iowa*, 210, 300, *300–303*; *Missouri*, 453–54, *453–56*; *New Jersey*, 355, 371, 373, *374–77*; *Wisconsin*, 210, 512, *513*
Mutsu-class battlecruisers (Japan), 3, 4
Myers, Charles E., 492
Myoko (Japan), 249

Nagasaki, 254, 403
Nagato (Japan), 4, *5*, *248*, 335, *336*, 406
Nagato-class battlecruisers (Japan), 3, 4
Nakajima B6N Tenzan Jill. *See* Japanese aircraft
National Defense Authorization Act (1996), 512

NATO: *Iowa* and, 261, 264, 275, *279–81*, 284; *New Jersey* and, 348; *Wisconsin* and, 264, 480, 486, 488
Nauticus National Maritime Center, Norfolk, Virginia, 512, *513*
navigation bridge, 98
navigation radar, 114
Navy Blue 5-N paint, 184
Navy Construction Act (1938), 13
Navy Day (1945), 411, *413*
Nevada-class battleships, 3, *8*
New Hampshire, 15
New Jersey: approval of, 15; armaments and armament modifications, 46–47, *46*; battle honors, *198*, *199*; commanding officers, *194*; commissioning, 304, *310*; construction, 21, 55, *304–6*, *308*, *309*, *311*; decommissioning, 348, 371, *372*; design complement, *163*; dimensions, displacement, and protection, 19; engine room control panel, *87*; enlisted men's canteen, *160*; forecastle teak after seventy-two years, *24*; 40-mm guns in action (1953), *70*; forward 16-inch turrets, *52*; forward superstructure of, *102*; Japanese surrender and after, 335–36; kamikaze attacks, *326*; during Korean War, *216*; launching, *307*, *308–9*; Marianas campaign, 323; Marshall Islands campaign, 243, 244, 316–17; metacentric height, 19; Mk-6 catapults and crane on, *133*; Mk-52 rangefinder in, *124*; modernization of, 267; as museum, 355, 371, 373, *374–77*; NATO and, 348; Operation Desecrate I, *245*, 319–20; Operation Hailstone, 240; Pacific War, *321–22*, *328*, *336*; paint and camouflage, *185*; Phalanx mount on board, *74*; plans for, *40–45*; previous ship named, *203*, 204; profile views (1983), *38–39*; at Puget Sound Naval Shipyard, 329, *331–35*, *355*; radars, *117*; radio call signals, 115; Reagan administration and recommissioning, **355**, *356–70*, 360–61, 366, 492; in reserve, 337; at sea, *96–97*; at sea in the 1950s, 338, *339–47*, 340–42, 347–48; service career, 304–81; shakedown cruise, 312; speed of, 82; SRBOC Mk-36 launchers on, *79*; steam trials, *313*; stereoscopic Mk-48 GFCS Mods 1–5 in, 123; Task Force 38 and, 324, *325*; Task Group 58.3 and, 328; Task Group 58.7 and, 328–29, 470; Tomahawk fired by, *77*, *214–15*; as training vessel for midshipmen, 336; Truk campaign, *318*, 321; typhoon and, *326–27*, 327; Vietnam War, 349, *349–53*, 352–53
New Mexico, 8, 15, 407
New Mexico-class battleships, 3
New Orleans, 238, 240, 245, 317, 466
New York, 7, 413
New York Naval Shipyard: history of, 15; *Iowa*, 15, 219, *223*; *Missouri*, 15, 382, 388–89; *New Jersey*, 337, 338; *New York*, 7; *South Dakota*, 9; *Wisconsin*, 470, 481
New York-class battleships, 3
Newport News Shipbuilding & Drydock Company, 205
Nimitz, Chester W., 258, 406
Nitro, 79
North, James R., *194*
North Carolina, 10, 11, 238, 245, 321, 466, 470
North Carolina-class battleships, 11
North Dakota, 6
Northern Wedding, Operation, 284, *285*

Nowaki (Japan), 240, *240*, 317, 324
nuclear warheads, 76

Oakland, 249, 324, *407*
Ocean Gray 5-0 paint, 184
O'Donnell, Edward, *194*
OE-82C/WSC-1 aerials, 115, *120*
officers' accommodations, *162*, 163, *164*, *165*
Ogden, Darin Andrew, *291*
Okinawa campaign, *58*, 82, 395, 398
operating theater, 180, *182–83*
Operations Center, *131*. *See also* plotting rooms
Osterwind, Robert, 340
Oyodo (Japan), 245
Ozawa Jisaburō, 245, 323, 324

Pacific Battleship Center (PBC), 210, 300, *300–303*. *See also Iowa*
Pacific War: battleships, escort carriers, heavy and light cruisers, and destroyers, *407*; *Iowa*, *237–39*, 238, 240, *241–50*, 242–45, 247, 249, 251, 253–54, *254*; Japanese advance, *378–79*; *Missouri*, 394, *396–412*, 398–99, 403, 406; *New Jersey*, 316–17, *318–19*, 319–21, *321–22*, 323–24, *325–29*, 327–29, 335–36, *336*; *Wisconsin*, 466, 468, *468–73*, 470–71, 473–74
paint and camouflage: *Iowa*, 188–89; *Missouri*, 190–91; paint shades and reflection factors, *185*; schematic drawings for, *186–87*; schemes, 184
Palau, raid on, 244
Panama Canal, 14, *273–74*, *351*, *368*
parallax corrector, 130
Pasadena, 407
Pearl Harbor (movie), 439
Pearl Harbor attack, 422, 452, *452*
Peckham, George E., *194*
Peng Dehuai, 430
Peniston, Robert, *194*, 492
Pennsylvania-class battleships, 3, *8*
periscopes, 98
Persecution, Operation, 320
Persian Gulf conflicts: *Iowa* and, 284; *Missouri* and, *437–50*, 439, 446; *New Jersey* and, 366; *Wisconsin* and, 505, *505–11*, 511
personnel boats, 28-foot, 153
Peterson, Ricky Ronald, *291*
petty officers' mess, *166*
Phalanx close-in weapon system (CIWS) guns, 355, 437, 498
Phalanx Mk-15 Block 0 CIWS, *46*, 74, *74–75*, 114
pharmacy, 180, *182*
Philadelphia, 16
Philadelphia Naval Shipyard: aerial view, *15*; history of, 16; *Iowa*-class battleships in reserve at, 491, *512*; *New Jersey* towed to, 349, 371; ship construction, 15, *22*, 209, *213*, 304, *305*, *308*, 312, 457, *461*; *Wisconsin* towed to, 488, *489–90*
Philippines, Pacific War and, 247, 324
Piasecki HUP-2 Retriever helicopter, *133*
Pick, Carl, 457
pilothouse, 98
Pittsburgh, 394, 399, 468, 470, 471, *473*
Pivot, Operation Red, 486
plotting rooms, *130*, *131*. *See also* battery plotting rooms; Operations Center
Plutón, 202
Polaris missiles, 209
powder: hatches for, *58*; smokeless diphenylamine, 47

Powell, Joseph W., 14
power and propulsion: engine plant, 82–83, *82–87, 92*; machine shop, *90*; propellers, *83, 93, 94–95*; rudders, *93, 95*; turbogenerators for electric motors, *88*; UNREP system, *91*
Price, Mathew Ray, 291
Princeton, 245, 249, 261
propellers, *82, 83, 93, 94–95*
provisions, 158, *158–62*
Puget Sound Naval Shipyard, Bremerton, Oregon, 329, *331–35*, 355, 434, 454
Pyotr Velikiy (USSR), 492–93

Quincy, *13*, 353, 407

radars and electronics: aerials on after stack, *121*; foremast, *115, 121*; forward superstructure, *118*; *Iowa*'s discone/discage antenna, *121*; *Missouri* refitting and, 437; *New Jersey* refitting and, 355; radar room, *122*; Satcom aerial, *120*; Satcom and ECM equipment, *119*; SK-2 air-search radar on *Missouri*, *119*; types and characteristics of, 114–15; types for each ship, *117*
Radford, Arthur W., 258, 328, 340, 471
radio call signals, 115
radiology services, 180, *181*
Randolph, 336, 470, 474, 486
range of fire, *Iowa* and record for, 47
rangefinders: armored on control tower, 98; battery plotting rooms and, *130–31*; data transmitted to plotting rooms, *130*; on *Iowa* (1950s), *128–29*; Mk-8 radar, *132*; Mk-37, *125, 126, 127*; Mk-41 gyroscope, *132*; Mk-51, *132*; Operations Center (1984), *131*; optics of, *124*; schematic drawings, *123*; types and characteristics of, 123
Ranger, *13*, 360, 439
Raymond, William H., 478
Raytheon, 76
Reagan, Ronald and administration, 266, 283, *283*, 355, 358, 360–61, 437, 492
Reeves, John W., 323
Reeves, Joseph, 11
refitting and conversion projects, 200–201
Regulus II launcher, 200
remote-controlled target ship, 202
remotely piloted vehicles (RPVs). *See* RQ-2B Pioneer remotely piloted vehicle
RGM-84 Harpoon antiship missiles, *46, 78, 78*
Rhee, Syngman, 342, 430, 478
Ridgway, Matthew B., 341, 480
Roosevelt, Franklin D., 231, *233*, 234–36, *237*; bathtub for, *165*, 300
Roosevelt, Theodore, 204
Roper, John W., 195, 336
Roremus, Irving S., 312
RQ-2B Pioneer remotely piloted vehicle (RPV), *133, 133*, 144–45, 284, 504
rudders, *82, 82, 93*
Russell, Guy, 434
Russell, Richard, 349
Ryuho (Japan), 470

SACEX (Supporting Arms Coordination Exercise), Operation, 284
Sacramento, 209, 449
Santiago, Battle of, 202
Saratoga, *4, 10, 13, 296*, 486, 488
satellite communications (Satcom), 115, *119, 120*
satellite launchers, 201

SCB 19, 209
SCB 173, 200
Schelin, Geoffrey Scott, 291
Schroeff, Coenraad, 195, 511
Scott lamps, signaling with, *147*
Scott-Montcrieff, Alan, 478
Seaquist, Larry, 192, 283, *293*
searchlights, 146, *147*
Second London Naval Conference (1936), 11
Second Vinson Act (1938), 13
secondary fire-control position, 98
secondary 5-inch armaments. *See* 5-inch guns
Secretary of the Navy: *Iowa*-class preservation and legal status and, 210. *See also specific Secretaries*
SG surface-search radar, 114, *117*
SG-6 surface-search radar, 114, *117*
Shafroth, John F., 398, 470
shaft revolution indicator, *105*
shafts, engine room, 82
Shangri-La, 253, *253*
shells: armor piercing (AP), colorant to distinguish, 47; 5-inch, 64, *66*; on the forecastle, *53*; heavy, awaiting stowage, *282*; 16-inch, *46, 48, 54, 58*, 340; 20-mm, 71
Sherman, Forrest P., 238, 316, 328
Shigemitsu Mamoru, 406, *409*
shipboard equipment: complement, 163, *163–79*; control tower and main directors, 98, *99–113*; embarked aviation, 133, *133–45*; ground tackle, 149, *149–52*; medical department, 180, *180–83*; provisions, 158, *158–62*; radars and electronics, 114–15, *114–22*; rangefinders, 123, *123–32*, 130; searchlights, 146, *146–47*; ship's boats, 153–54, *153–57*
ship's log, 98
shoe cobblers, *419*
Shō-Ichi-Gō ("Victory"), Operation, 247, 324
Shonan Maru No. 15 (Japan), 240, 317
side-wheel frigate named *Missouri*, 204
side-wheel ironclad steamer named *Missouri* in Confederate Navy, 204
signal locker, *112*
Sikorsky CH-53 Sea Stallion (helicopter), 133
Sikorsky CH-53E Super Stallion (helicopter), 133, *143*
Sikorsky HO3S-1 Horse (helicopter), 133, *138*
Sikorsky SH-3A Sea King (helicopter), 133
Sikorsky SH-60 LAMPS MK III Seahawk (helicopter), 133
Sikorsky SH-60B Seahawk (helicopter), *142*
16-inch guns: characteristics and protection of, *48*; civilians during shoot of, *55*; inspecting breach of, on *Wisconsin*, *58*; *Iowa*-class, saving from demolition, 437; for *Iowa*-class ships, 14, 46–47, *46*; loading, *293*; loading phases, *59–61*; *North Carolina*-class designs and, 11, 12; radars for, 114; violence of, *247*
SK air-search radar, 114, *117*
SK-2 air-search radar, 114, *117, 118, 119*
Skinner, W. Mark, 300
SLQ-32 (V)3 ECM, *119*
Smedberg, William, 192, 258, 261
Smith, Allan E., 422, 428
Smith, Harold P., 194, 426
smokeless powder diphenylamine (SPD), 47
Snyder, Edward, 194, 349, 492
Sohn Won-yil, 430, 478
Solomons, Edward W., 192
South Carolina–class battleships, 3

South Dakota: basic specifications, *12*; construction, 9; Halsey's flag shifted to, 411; Japanese surrender and, 407; Pacific War, 245, 249, 321, *322*, 328, 398, 466, 470
South Dakota–class battleships: cancellation of, 4, 202; *Iowa*-class designs and, 14; Pacific War, *325*; 16-inch barrel for, *5*; Washington Treaty and, 12–13
SP high-altitude search radar, 114, *117*
Spanish-American War (1898), 3, 202
speeds, *Iowa*-class, *82, 82*
Spellman, Francis, 478
Spence, 245, 251, 327
spots 1 and 2, for rangefinders, 123
Sprague, Clifton A., 258
Springboard, Operation, 481
Springfield, 407
Spruance, Raymond A., 202, 238, 317, 319, 320, 394
SPS-4 surface-search radar, 114, *117*
SPS-6 air-search radar, 114, *115, 117*, 216
SPS-8 high-altitude search radar, 114, *117*
SPS-10 navigation radar, 114, *117*
SPS-10 surface-search radar, 114
SPS-12 air-search radar, 114, *115, 117*
SPS-49 air-search radar, 114, *117*
SPS-53 surface-search radar, 114
SPS-67 surface-search radar, 114, *117*
SQ surface-search radar, 114, *117*
SR air-search radar, 114, *117*
SR-3 air-search radar, 114, *117*
SRBOC Mk-36 (Super Rapid Bloom Offboard Chaff) countermeasures system, 79; launchers for, *75, 79*
Standley, William, 12
Statue of Liberty centennial, 283, *283*
steam plant, 82. *See also* engine plant
Stefan, David R., 328
Stillwagon, Heath Eugene, 291
Stirling, Yates, 205
Stokes, Thomas M., 192
Stone, Earl E., 195, 457
Strauss, Joseph, 14
Strikeback, Operation, 264, 488, *489*
Struble, Arthur D., 428
SU surface-search radar, 114, *117*
Sullivan, John, 420
Swanson, Claude A., 11, 12
Sylvania, 336
Sylvester, John, 194

Talos missile launchers, 200, 209
Tama (Japan), 245
Tampico Affair, 204
Tan No. 2, Operation, 470
Tarbuck, Raymond D., 192
targets, inflatable, for gunnery exercises (1980s), *62*
Tartar missile launchers, 200, 209
Task Force 38 and, 247, 324, *325*, 399, 403, 466, 471
Task Force 58: *Iowa* assigned to, 238, 244; *Missouri* and, 394; *New Jersey* and, 244; Operation Desecrate I and, 245, 246, 319–20; *Wisconsin* and, 468; *Yamato* destroyed by, 398, 471
Task Force 77, 340–41, 478
Task Force 81, 336, 337, 474
Task Force 95, 428
Task Group 27.5, 234, 236
Task Group 30.1, 399

Task Group 38.2, 247, 249, 324, 327
Task Group 38.3, 249
Task Group 38.4, 253, 399, 403
Task Group 50.9, 238, 240
Task Group 58.1, 395
Task Group 58.2, 319, 321, 394, 395, 468
Task Group 58.3, 238, 243, 316, 319, 323, 328
Task Group 58.4, 245, 247, 323, 328, 395, 471
Task Group 58.7, 328, 398, 470–71
Task Group 70.10, 439
Task Unit 28.1.1, *196*
TASM (Tomahawk Anti-Ship Missile), 76
Tatham, Todd Thomas, *291*
technical characteristics: armor belts, 23, *23*; armor deck hatch, *25*; builders of, 15–16; decks, 23; dimensions, displacement, and protection, 19; genesis of, 3–11; Itō Seiichi, *20*; metacentric height, 19; plans for, *27, 32–33, 40–45*; profile views, *38–39*; Washington Naval Conference agreement and, 11–15
technical records office, *129*
Teledyne J402 turbojet engine: for Harpoon missiles, 78
telephones, 83
telescopes: Mk-66, in turrets, 123; sighting, in Mk-38 directors, 123
Tennessee: built at New York Naval Shipyard, 15
TERCOM (TERrain COntour Matching) inertial navigation system: for Tomahawk missiles, 114
Terrier antiaircraft armament, 209
Thach, James, *194*
Thompson, 261
Thompson, Edward, *194*, 335
Thompson, Jack Ernest, *291*
Thompson, Thomas G., 498, 511
three-gun turrets, 46. *See also* turrets
Thurber, H. Raymond, 341
Ticonderoga, 466; Task Group 38.3 and, 249; Task Group 38.4 and, 253
Tirpitz, 231, *233*
TLAM-C (Tomahawk Land-Attack Missile Conventional), 76
TLAM-N (Tomahawk Land-Attack Missile Nuclear), 76
Tomahawk cruise missiles, *46*, 76–77
Tomahawk system: *Iowa* and, 267, *286*; *Missouri* and, 437; *New Jersey* and, 355, *358*; pre-launch programming for, 114; types and components, 76, *76*; *Wisconsin* and, 498
Tower, John, 498
Towers, John H., 335–36
Toyoda Soemu, 247
training ship: *Iowa* as, 261, 263, 294, *298*; *Missouri* as, 415, 426, 434; *New Jersey* as, 336, 341, 348; previous *Missouri* as, 205; previous *New Jersey* as, 204; previous *Wisconsin* as, 205; *Wisconsin* as, 478, 486, 488
Trippe, 91
Truk campaign, 240, *241*, 242
Truman, Harry S., 382, *386*, 406, 412, 414, 415, 420
Truman, Margaret, 382, *386*, 414, 437

tubes, armored, for transmitting orders, 98
Tucker, Ronald, *194*, 366
turbines, 82, *82*
turbogenerators for electric motors, 82, *88*
turrets: *Iowa*'s, *56–57*, 286; Mk-52 coincidence rangefinders in, 123; painting stars and stripes on, *178*; plans for, *50*; 16-inch Mk-7, 46–47, *50*, 52; 16-inch Mk-7, access to, *56*; telescopic sights on, *62*
20-mm 70-caliber Mk-4 guns, 71, 229, *242*
Type-P Mk-6 catapults, 133
Typhoon Cobra (Halsey's Typhoon), 251
typhoons: Korean War, 428; Pacific War, *326–27*, 327, 399, 468, *468*, 471
Tyree, David, *194*, 338

Umezu Yoshijirō, 406
Under Siege (movie), 439
underway replenishment (UNREP) system, 83, 91, 428–29
United States, Washington Naval Conference agreement and, 3–4, *4*
utility boat (UB), 40-foot Mk-3, 154

Vanguard, HMS, 83, 480
Varley, Richard, 481
Vasey, Lloyd, 353
vertical missile launcher, 200
Vietnam War, 349, *349–53*, 352–53, *381*
Vincennes, 249
Virginia-class battleships, *203*, 204
Vizcaya, 202
Vought OS2U Kingfisher reconnaissance and antisubmarine floatplanes: on catapult, *395*; as embarked aircraft, 133; launch trials, *312*; launching, *135, 136, 250*; mission to Guam, 323; on *New Jersey*, *314*; specifications, *134*
VPS-2 Pulse Doppler radar, 114

W80 nuclear warhead, for Tomahawk missiles, 76
Wallace, Ilo, 219, *225*
Wallin, Homer N., 422, *422, 427*
Walter, Wilfred A., 235
Washington: approval of, 11; construction, 16; Lee and, 323; off Port Angeles, Washington (1944), *11*; Task Force 38 and, 466; Task Force 58 and, 238, *246*; Task Group 58.7 and, 328, 398, 470
Washington Naval Conference and Treaty (1921–22), 3–4, 11, 14
Wasp, 13, 395, 466, 470, 480, 488
water, drinking or running, 158
watertight doors, *146*
Way, David, *300*
Weinberger, Caspar W., 360, 437
Welden, Stephen J., *291*
Wellborn, Charles, Jr., *192*, 253, 348
Westinghouse reduction gear, 82, *82*
whaleboats, 26-foot Mk-10, 154
whaleboats, 26-foot motor, 153, 154
White, James Darrell, *291*
White, Rodney Maurice, *291*
Wichita, 16, 238, *245*, 316

Wilkes Barre, 328, *407*
William Cramp & Sons Shipbuilding Company, Philadelphia, 202
William D. Porter, 234–35, *236*
Williams, Michael Robert, *291*
Wilson, 238
Wilson, Pete, 437
Windham Bay, 399, 471
windlasses, anchor, 149, *152*
Wisconsin: approval of, 15; armaments and armament modifications, 46–47, *46*; battle honors, *198*; collision with *Eaton* and after, 481, *482–84*; commanding officers, *195*; commissioning, 465; construction, 21, *22*, 195, 457, *458–60*; decommissioning, 474, 491, 512, *512*; design complement, 163; dimensions, displacement, and protection, 19; enlisted men's quarters, *167*; fitting out, *462–63*; Gulf War, 505, *505–11*, 511; at Hampton Roads Naval Review (1957), *486*; handling powder bags during Okinawa campaign, 58; inspecting breach of 16-inch gun in, 58; Korean War, 341, 342, 475, *475–77*, 478, *479–80*; launching, *461*; life aboard, *464–65*, 466; Mk-38 Mods 6–7 rangefinders in, 123; as museum, 210, 512, *513*; NATO and, 264, 480, 486, 488; Pacific War, 466, 468, *468–73*, 470–71, 473–74; paint and camouflage, *185*; Persian Gulf conflicts, 446; at Philadelphia in reserve, *490–91*; previous ship named, 205, *206*, 457; radars, *117*; radio call signals, 115; Reagan administration and recommissioning, 492, *494–504*, 498–99; at sea in 1998, *26*; service career, 457–513; 16-inch turrets, *52*; steam trials, 466, *467*; Tomahawk fired by, *77*
Woodward, Clark H., *20*, 219
Wooldridge, Edmund, *194*, 327
Woolfidge, E. T., 261
Wright, George T., *194*
Wyoming, 7
Wyoming-class battleships, 3

X-ray room, *181*

Yahagi (Japan), 328, 471
Yamato (Japan): destruction of, 470, *470*, 471; fitting out at Kure (1941), *14*; before Operation Shō-Ichi-Gō, *248*; portrait of, *329*; sophisticated protection for, 249; suicide mission by, 328–29, 398
Yamato-class battleships (Japan), 14
Yokosuka D4Y Judy dive bombers, 249, 328
Yorktown: battleship speeds compared with speed of, 13; kamikaze attacks, 395; Pacific War, 240, *245*, 253, 328, 403, 466, 471
Young, 234
Young, John Rodney, *291*

Ziegler, Reginald Owen, *291*
Zuiho (Japan), 245
Zuikaku (Japan), 245
Zuni Folding-Fin Aircraft Rockets (FFAR), 5-inch, *46*

About the Author

Philippe Caresse was born into a naval family in France in 1964 and joined the Marine Nationale in 1982, serving in the destroyer *d'Estrées*. He has published an extensive range of ship monographs on the French, German, U.S., and Japanese navies from the late nineteenth century to World War II and is coauthor with John Jordan of *French Battleships of World War One* (2017) and the forthcoming *French Armoured Cruisers*. A diving instructor, he is the harbormaster of a marina on the Côte d'Azur.

The Naval Institute Press is the book-publishing arm of the U.S. Naval Institute, a private, nonprofit, membership society for sea service professionals and others who share an interest in naval and maritime affairs. Established in 1873 at the U.S. Naval Academy in Annapolis, Maryland, where its offices remain today, the Naval Institute has members worldwide.

Members of the Naval Institute support the education programs of the society and receive the influential monthly magazine *Proceedings* or the colorful bimonthly magazine *Naval History* and discounts on fine nautical prints and on ship and aircraft photos. They also have access to the transcripts of the Institute's Oral History Program and get discounted admission to any of the Institute-sponsored seminars offered around the country.

The Naval Institute's book-publishing program, begun in 1898 with basic guides to naval practices, has broadened its scope to include books of more general interest. Now the Naval Institute Press publishes about seventy titles each year, ranging from how-to books on boating and navigation to battle histories, biographies, ship and aircraft guides, and novels. Institute members receive significant discounts on the Press' more than eight hundred books in print.

Full-time students are eligible for special half-price membership rates. Life memberships are also available.

For a free catalog describing Naval Institute Press books currently available, and for further information about joining the U.S. Naval Institute, please write to:

Member Services
U.S. NAVAL INSTITUTE
291 Wood Road
Annapolis, MD 21402-5034
Telephone: (800) 233-8764
Fax: (410) 571-1703
Web address: www.usni.org